Biomedical Imaging Technologies for Cardiovascular Disease

Biomedical Imaging Technologies for Cardiovascular Disease

Editor

Julio Garcia Flores

MDPI • Basel • Beijing • Wuhan • Barcelona • Belgrade • Manchester • Tokyo • Cluj • Tianjin

Editor
Julio Garcia Flores
University of Calgary
Canada

Editorial Office
MDPI
St. Alban-Anlage 66
4052 Basel, Switzerland

This is a reprint of articles from the Special Issue published online in the open access journal *Applied Sciences* (ISSN 2076-3417) (available at: https://www.mdpi.com/journal/applsci/special_issues/Imaging_Cardiovascular).

For citation purposes, cite each article independently as indicated on the article page online and as indicated below:

LastName, A.A.; LastName, B.B.; LastName, C.C. Article Title. *Journal Name* **Year**, *Volume Number*, Page Range.

ISBN 978-3-0365-7006-8 (Hbk)
ISBN 978-3-0365-7007-5 (PDF)

Contents

About the Editor

Julio Garcia Flores

Julio Garcia Flores, Ph.D. is an Assistant Professor within the Department of Radiology at the University of Calgary. His research focuses on cardiovascular imaging to assess heart valve disease, aortopathies, and congenital diseases. He leads the pre-clinical research program at the Stephenson Cardiac Imaging Centre where he directs the larger 4D-Flow program in Canada. He is a Fellow of the Mexican National Researcher's System, Junior Fellow of the International Society for Magnetic Resonance in Medicine, Senior Member of the Institute of Electrical and Electronics Engineers, and Fellow of the Society for Cardiovascular Magnetic Resonance. Dr. Garcia's publications includes > 180 works including original articles and conferences with >2000 citations. He was performed over 50 international lectures and obtained over 40 international awards.

Preface to "Biomedical Imaging Technologies for Cardiovascular Disease"

This Special Issue in *Biomedical Imaging Technologies for Cardiovascular Disease* belongs to the field of Applied Biosciences and Bioengineering within Applied Sciences. Readers will discover some of the most recent innovations in cardiovascular imaging. As these novel imaging applications become increasingly mainstream for clinical and research applications, it is important for physicians and imaging specialists to be up to date on their appropriate use and added value. The included articles can be classified into the following three topics: I) Advances in Doppler Echocardiography; II) Advances in Magnetic Resonance Imaging; and III) Specialized Reviews. The reader will learn vital information about the fundamental imaging assessment of heart failure, therapeutic cross-link strategies for cardiovascular disease, pediatric Doppler imaging of pulmonary hypertension, machine learning applications for left ventricle hypertrophy, and the development and clinical applications of three-dimensional time-resolve blood flow imaging by magnetic resonance imaging (4D-Flow MRI). Furthermore, the Special Issue includes two comprehensive reviews on coronary disease imaging using positron emission tomography and on the emerging use of ultrasmall supermagnetic particles in cardiac imaging. The editorial in Chapter provides a more comprehensive summary of each article.

Julio Garcia Flores
Editor

applied sciences

MDPI

Editorial

Biomedical Imaging Technologies for Cardiovascular Disease

Julio Garcia [1,2,3,4,5]

1 Department of Radiology, Cumming School of Medicine, University of Calgary, Calgary, AB T2N 1N4, Canada; julio.garciaflores@ucalgary.ca
2 Department of Cardiac Sciences, Cumming School of Medicine, University of Calgary, Calgary, AB T2N 1N4, Canada
3 Libin Cardiovascular Institute, University of Calgary, Calgary, AB T2N 1N4, Canada
4 Alberta Children's Hospital Research Institute, University of Calgary, Calgary, AB T2N 1N4, Canada
5 Stephenson Cardiac Imaging Centre, University of Calgary, Calgary, AB T2N 1N4, Canada

1. Introduction

Non-invasive biomedical imaging technologies for investigating the heart's physiology, performance, function, and structure have increased in number and diversity over the past several years. The standard-of-care imaging techniques include echocardiography, X-rays, computed tomography, nuclear scans, magnetic resonance imaging (MRI), and catheterization. Each imaging technique provides specific information, with its own characteristics, advantages, and limitations. Important improvements in precision, sensitivity, accuracy, and refinement have been possible thanks to the technological advances in software and hardware. Better assessments of cardiovascular disease, hemodynamics, and biomarkers have aided the stratification of patient risk and therapy. The recent integration of artificial intelligence and machine learning have also supported novel approaches for personalized image-based diagnosis. Furthermore, image-based computation-assisted diagnosis is a promising means of uncovering key insights into disease progression and personalized solutions.

This Special Issue on Biomedical Imaging Technologies for Cardiovascular Disease collected the most recent progress in biomedical imaging technologies for cardiovascular disease. The original articles and reviews cover a wide range of subjects that can be classified into the following topics: (I) advances in Doppler echocardiography; (II) advances in magnetic resonance imaging; (III) specialized reviews.

2. Topic I: Advances in Doppler Echocardiography

Doppler echocardiography remains the primary imaging technique for the assessment of cardiovascular disease. This non-invasive imaging technique has a low cost, is easily accessible, and can be used in standard-of-care primary cardiovascular evaluations. The classification and monitoring progression proposed by the current AHA/ACC includes four stages that consider the factors that influence clinical decision making: severity status, irreversible consequences in the left and right ventricles, structural and functional changes in the whole heart, changes in cardiac pressures, development of cardiac arrythmias, and the presence of symptoms.

Heart failure (HF) is the most common cardiovascular disease worldwide affecting > 25 million people in industrialized countries [1]. The most widely used device-based therapy for the treatment of heart failure with a reduced ejection fraction (HFrEF) and a QRS duration > 150 ms is cardiac resynchronization therapy. Cardiac contractility modulation (CCM) has been available for patients with a narrow QRS and persistent symptoms or frequent HF hospitalization re-admissions. CCM improves the cardiac performance and reverses left ventricular (LV) remodeling. In this Special Issue, Contaldi et al. [2] analyzed the effects of CCM on right ventricular (RV) systolic function and RV–pulmonary artery (PA) coupling. They reported that HF patients who underwent CCM therapy showed a reduction of the RV diameters and improved RV systolic function, PA systolic pressure, and

Citation: Garcia, J. Biomedical Imaging Technologies for Cardiovascular Disease. *Appl. Sci.* **2023**, *13*, 2209. https://doi.org/10.3390/app13042209

Received: 4 February 2023
Accepted: 7 February 2023
Published: 9 February 2023

RV–PA coupling after six months of follow-up. The latter findings remark on the potential benefit of CCM as device-based therapy for HF patients.

The therapeutic cross-link between urology and cardiovascular disease has been well reported and recognized. For example, the inhibitors of phosphodiesterase type 5 (PDE5i) are a common treatment for erectile dysfunction and are used to treat pulmonary hypertension. PDE5i holds back the breakdown of nitric oxide-driven cyclic guanosine monophosphate (cGMP) in smooth cells of the vascular bed, resulting in a potent vasodilator. cGMP signaling is abnormal in HF, and it has emerged as potential therapeutic option [3]. Crocetto et al. investigated whether tadalafil, a long-acting PDE5i, could also improve the diastolic function as assessed by echocardiography [4]. A total of 23 patients underwent the echocardiography assessment at the baseline, after 3 months, and after 6 months. This prospective study showed that after 6 months of erectile dysfunction therapy for secondary to radial prostatectomy, a positive effect was observed for the diastolic function by improving the E/e′ ration and peak velocity of the tricuspid jet. These findings suggest that a further investigation could be performed to assess the role of tadalafil as a treatment and to prevent diastolic dysfunction.

Tissue motion annular displacement (TMAD) is a well-known speckle-tracking-derived echocardiographic parameter that tracks the annular tissue towards the ventricular apex. TMAD is an angle-independent parameter that is commonly measured at the mitral and tricuspid valves and used for the assessment of ventricular function. In this Special Issue, Muntean et al. investigated the prognostic value of TMAD of the tricuspid valve in children with pulmonary arterial hypertension (PAH) [5]. PAH is a severe and progressive disease that causes RV disfunction over time in children. Its timely characterization and detection can improve pediatric PAH treatments. In this study, twenty PAH patient and twenty sex- and age-matched controls were enrolled at the baseline and after one year of follow-up. By comparing the healthy and poor PAH groups with the controls, the TMAD indices were significantly lower in the PAH groups ($p = 0.010$ and $p = 0.018$, respectively). These findings suggest that TMAD could be a good parameter to include in the assessment of pediatric PAH patients.

3. Topic II: Advances in Magnetic Resonance Imaging

The magnetic field properties of non-invasive MRIs cause the body's hydrogen atoms to align temporarily. Radiofrequency pulses are then applied to stimulate the protons out of equilibrium. The frequential response signal is collected and used to reconstruct an image of the region of interest. Cardiovascular MRIs provide detailed images to examine the heart's function, structure, cardiac motion, and tissue damage of the cardiac muscle. An additional advantage is the lack of radiation and use of iodine contrast.

Left ventricular hypertrophy (LVH) is presented in between 15% and 20% of the population. LVH is more frequent in patients with hypertension and obesity and in Afro-American people. Cardiac MRIs provide a comprehensive assessment of myocardial hypertrophy. Cardiac MRI has become the gold standard for ventricular volume quantification. Accurate measurements of the wall thickness, hypertrophy pattern, chamber size, and ventricular function are necessary for the proper characterization of LVH. These measurements often require the delineation of the region of interest in a large number of images per patient. In recent years, machine learning (ML) methods have been used to accelerate the segmentation of the ventricles for LV function quantification. In this Special Issue, Budai et al. aimed to automatically detect LVH in cardiac MRI scans by applying ML methods [6]. The authors developed an ensemble model based on a 3D ResNet. The study included 428 patients with LVH under various conditions and 234 healthy controls without any known cardiovascular diseases. The developed model achieved a 92% F1 score and 97% recall using the testing dataset, which are similar to those of expert reports. This study is an excellent example of the clinical cardiac applications of ML.

Standard-of-care cardiac MRI protocols typically use 2D phase contrast (PC) to assess heart hemodynamics. However, 2D PC only acquires the through-plane velocity for the flow quantification in the region/vessel of interest. A whole heart 3D hemodynamic acquisition of the velocity field can be achieved using a 4D-flow MRI. This Special Issue collected four studies investigating the clinical applications, validation, and novel approaches for 4D-flow MRI. The study by Hong et al. [7] aimed to investigate the hemodynamic changes within the PA and its association with ventricular remodeling and vascular architecture. This study included 33 patients with PAH and 17 healthy controls. The patients presented alterations in the PA, showing larger distances of PA bifurcation. In addition, the wall shear stress (WSS), kinetic energy (KE), and energy loss (EL) values decreased further in the PAH patients compared to those of the controls. The PA 4D-flow derived velocities were associated with RV ejection fraction. This study illustrated the potential usefulness of 4D-flow MRI for the hemodynamic assessment of the right heart, which is often more challenging to assess than the left heart.

One important aspect in the 4D-flow acquisition planning is the subject's respiratory motion. Respiratory gating is typically used to reduce the effects of respiratory motion and artifacts. Denecken et al. [8] performed a comparison of 4D-flow acquisitions with and without respiratory self-gating and its impact on the 4D-flow derived parameters. This study was conducted with 15 healthy controls. Their findings showed that the acquisition without respiratory self-gating underestimated the values in the aortic arch, the descending aorta, and diaphragmatic aorta. Both the acquisitions with and without respiratory self-gating showed a significant variability in the 4D-flow-derived parameters. However, respiratory compensation provided a more reliable quantification.

Advanced cardiac hemodynamic characterization ideally requires high temporal and spatial resolution datasets. A standard 4D-flow MRI acquisition can have >4000 images with an isotropic spatial resolution of 2–3 mm per voxel and 25–45 ms of temporal resolution. Despite the impressive characteristics of the standard 4D-flow acquisition, given the multidimensional and complex nature of the cardiac flow, it is difficult to assess the hemodynamic phenomena on a small scale. Recent studies applied finite element strategies to achieve computation fluid dynamics simulation characteristics. One application example is the study presented by Franco et al. [9]. In this study, the authors presented a proof of concept for the use of finite element methods in dilated cardiomyopathy (DCM). A total of 25 subjects, 12 controls, and 13 DCM patients were recruited to perform a comprehensive evaluation of the left intra-ventricular hemodynamics. Their approach demonstrated high inter- and intra-observer reproducibility and detected abnormal flow patterns in more detail than standard 4D-flow acquisitions can.

As cardiac MRI technology improves and 4D-flow MRI sequences become more widely used, several investigators have begun to question the inter- and intra-observer reproducibility in the whole heart. Juffermans et al. evaluated this aspect in the thoracic aorta for the most common 4D-flow derived parameters [10]. A cohort of 20 patients with aortic aneurysms were recruited for this purpose. They reported good reproducibility for normalized flow displacement and jet angle, a very good-to-excellent reproducibility for WSS and helicity, and an excellent reproducibility for vorticity.

These four studies in 4D-flow MRI illustrate the impressive technical and clinical application progress that has been made in recent years for this advanced cardiac flow technique.

4. Topic III: Specialized Reviews

This Special Issue included two review articles on positron emission tomography (PET) in coronary heart disease and the emerging use of ultrasmall supermagnetic particles iron oxide (USPIO) in cardiac MRI. De Almeida et al.'s review of PET imaging presented the recent advances in plaque imaging, and the use of 18F-Fluorodeoxyglucose for inflammation imaging, microcalcification imaging using 18F-Sodium Fluoride, myocardial perfusion imaging with Ribidium, 13N-Ammonia, and 15O-water, and viability imaging [11]. Tsampasian et al. presented a great summary of the properties of USPIO, its use in cardiac MRI, clinical applications, and safety parameters [12]. Both of these reviews assembled the state-of-the-art knowledge for PET imaging in coronary disease and USPIO in cardiovascular diseases.

Acknowledgments: This Special Issue would not be possible without the contribution of enthusiastic and talented authors. I would like to express my great gratitude to all of the reviewers. Your extraordinary feedback allowed the authors to improve their manuscripts and sustain the high-quality peer review standards. I would like to recognize the amazing work from the dedicated editorial team of *Applied Sciences* in assembling this Special Issue.

Conflicts of Interest: The author declares no conflict of interest.

References

1. Savarese, G.; Lund, L.H. Global Public Health Burden of Heart Failure. *Card. Fail. Rev.* **2017**, *3*, 7. [CrossRef] [PubMed]
2. Contaldi, C.; De Vivo, S.; Martucci, M.L.; D'Onofrio, A.; Ammendola, E.; Nigro, G.; Errigo, V.; Pacileo, G.; Masarone, D. Effects of Cardiac Contractility Modulation Therapy on Right Ventricular Function: An Echocardiographic Study. *Appl. Sci.* **2022**, *12*, 7917. [CrossRef]
3. Kong, Q.; Blanton, R.M. Protein Kinase GI and Heart Failure. *Circ. Heart Fail.* **2013**, *6*, 1268–1283. [CrossRef] [PubMed]
4. Crocetto, F.; Cuomo, V.; Santoro, C.; Fiorillo, L.; Barone, B.; Arcaniolo, D.; Fedele, T.; Imbimbo, C.; Esposito, R. New Possibilities in Heart Failure: The Effects of Tadalafil on Diastolic Function in Patients Undergoing Robot-Assisted0 Radical Prostatectomy. *Appl. Sci.* **2022**, *12*, 5629. [CrossRef]
5. Muntean, I.; Melinte, M.; Făgără, A.; Șuteu, C.C.; Togănel, R. A Novel Speckle-Tracking Echocardiography Derived Parameter That Predicts Clinical Worsening in Children with Pulmonary Arterial Hypertension. *Appl. Sci.* **2022**, *12*, 5494. [CrossRef]
6. Budai, A.; Suhai, F.I.; Csorba, K.; Dohy, Z.; Szabo, L.; Merkely, B.; Vago, H. Automated Classification of Left Ventricular Hypertrophy on Cardiac MRI. *Appl. Sci.* **2022**, *12*, 4151. [CrossRef]
7. Hong, Z.M.; Garcia, J. Pulmonary Artery Remodeling and Advanced Hemodynamics: Magnetic Resonance Imaging Biomarkers of Pulmonary Hypertension. *Appl. Sci.* **2022**, *12*, 3518. [CrossRef]
8. Denecken, E.; Sotelo, J.; Arrieta, C.; Andia, M.E.; Uribe, S. Impact of Respiratory Gating on Hemodynamic Parameters from 4D Flow MRI. *Appl. Sci.* **2022**, *12*, 2943. [CrossRef]
9. Franco, P.; Sotelo, J.; Montalba, C.; Ruijsink, B.; Kerfoot, E.; Nordsletten, D.; Mura, J.; Hurtado, D.; Uribe, S. Comprehensive Assessment of Left Intraventricular Hemodynamics Using a Finite Element Method: An Application to Dilated Cardiomyopathy Patients. *Appl. Sci.* **2021**, *11*, 11165. [CrossRef]
10. Juffermans, J.F.; van Assen, H.C.; te Kiefte, B.J.; Ramaekers, M.J.; van der Palen, R.L.; van den Boogaard, P.; Adriaans, B.P.; Wildberger, J.E.; Dekkers, I.A.; Scholte, A.J.; et al. 4D Flow MRI in Ascending Aortic Aneurysms: Reproducibility of Hemodynamic Parameters. *Appl. Sci.* **2022**, *12*, 3912. [CrossRef]
11. de Almeida, J.; Martinho, S.; Gonçalves, L.; Ferreira, M. Positron Emission Tomography in Coronary Heart Disease. *Appl. Sci.* **2022**, *12*, 4704. [CrossRef]
12. Tsampasian, V.; Merinopoulos, I.; Cameron, D.; Garg, P.; Vassiliou, V.S. Ultrasmall Superparamagnetic Particles of Iron Oxide and Cardiac Magnetic Resonance: Novel Imaging in Everyday Conditions. *Appl. Sci.* **2022**, *12*, 6913. [CrossRef]

applied sciences

MDPI

Article

Effects of Cardiac Contractility Modulation Therapy on Right Ventricular Function: An Echocardiographic Study

Carla Contaldi [1,*], Stefano De Vivo [2], Maria L. Martucci [1], Antonio D'Onofrio [2], Ernesto Ammendola [1], Gerardo Nigro [3], Vittoria Errigo [1], Giuseppe Pacileo [1] and Daniele Masarone [1]

1 Heart Failure Unit, Department of Cardiology, AORN dei Colli Monaldi Hospital, 80131 Naples, Italy
2 Electrophysiology Unit, Department of Cardiology, AORN dei Colli Monaldi Hospital, 80131 Naples, Italy
3 Cardiology Unit, Department of Medical Translational Sciences, University of Campania "Luigi Vanvitelli", 80131 Naples, Italy
* Correspondence: contaldi.carla@gmail.com

Abstract: Background. Cardiac contractility modulation (CCM) is a novel device-based therapy for patients with heart failure with reduced and mild reduced ejection fraction (HFrEF/HFmrEF). CCM increases cardiac performance and produces reverse left ventricular remodeling, with improved symptoms, functional capacity, quality of life, and reduced HF-related hospitalizations. However, to date, little evidence is available on the effects of CCM on right ventricle (RV) function. Therefore, we analyzed the effects of CCM on RV systolic function and RV–pulmonary artery (PA) coupling. **Methods.** Twenty-one (65 ± 12.5 years) patients with NYHA class III, ejection fraction < 40% and QRS < 120 ms were assessed at baseline. During follow up, two patients had died, and so nineteen patients were evaluated six months after CCM therapy. Using echocardiography, tricuspid annular systolic excursion (TAPSE), myocardial systolic excursion velocity (RVs), and RV free-wall strain was measured. PA systolic pressure (PASP) was estimated from tricuspid regurgitation, adding the right atrial pressure estimation. The RV-PA coupling was calculated as TAPSE/PASP ratio. **Results.** After six months, patients who underwent CCM therapy showed a reduction in RV diameters and improved RV systolic function, as evidenced by the increase in both TAPSE (16.6 ± 4.2 mm vs. 18.5 ± 3.6 mm; $p < 0.05$), RVs (10.1 ± 1.8 cm/s vs. 11.3 ± 11.4 cm/s; $p < 0.05$), and RV strain (−13.7 ± 1.8% vs. −15.6 ± 2.3%; $p < 0.05$). CCM also determined a reduction in PASP (34.2 ± 9.8 mmHg vs. 28 ± 6.2 mmHg; $p < 0.05$) and an increase in the TAPSE/PASP ratio (0.52 ± 0.14 mm/mmHg vs. 0.66 ± 0.23 mm/mmHg; $p < 0.05$). **Conclusions.** At six months, CCM increases RV reverse remodeling and performance, reducing RV size and improving RV systolic function, PASP, and RV-PA coupling.

Keywords: cardiac contractility modulation; right ventricular function; right ventricle–pulmonary artery coupling

Citation: Contaldi, C.; De Vivo, S.; Martucci, M.L.; D'Onofrio, A.; Ammendola, E.; Nigro, G.; Errigo, V.; Pacileo, G.; Masarone, D. Effects of Cardiac Contractility Modulation Therapy on Right Ventricular Function: An Echocardiographic Study. *Appl. Sci.* **2022**, *12*, 7917. https://doi.org/10.3390/app12157917

Academic Editor: Julio Garcia Flores

Received: 23 June 2022
Accepted: 5 August 2022
Published: 7 August 2022

1. Introduction

Heart failure (HF) is the most common cardiovascular disease worldwide, with more than 25 million people being affected in industrialized countries [1].

Despite advances in pharmacologic therapies for treating patients with HF with reduced ejection fraction (HFrEF), the prognosis of such patients remains poor [2,3]; therefore, device-based therapy has become increasingly important in recent years for the treatment of HFrEF.

The most widely used device-based therapy for the treatment of HFrEF is cardiac resynchronization therapy, which can lead to improved cardiac performance and prognosis in patients with HFrEF and wide QRS (duration > 150 ms) [4].

Unfortunately, only 30% of HFrEF have a QRS duration > 150 ms [5]. For patients with persistent symptoms or frequent HF-related hospitalizations but with narrow QRS, a new device (Optimizer Smart®) capable of delivering cardiac contractility modulation (CCM) therapy has been available for several years [6].

Optimizer Smart delivers non-excitatory electrical impulses on the interventricular septum during the absolute refractory period of cardiomyocytes; therefore, it does not alter the cellular action potential and does not affect heart rhythm [7].

CCM affects Ca^{2+} handling in cardiomyocytes, the expression of genes encoding factors that play a crucial role in Ca^{2+} metabolism, and molecular pathways involved in interstitial fibrosis [8,9].

These actions result in reverse left ventricular remodeling and increased cardiac performance, improving symptoms, functional capacity, quality of life, and reduced HF-related hospitalizations [10,11].

However, to date, there is little evidence available on the effect of CCM on right ventricular performance; thus, in this study, we analyzed the effects of CCM on right ventricular systolic function and right ventricle–pulmonary artery (RV-PA) coupling in patients with HFrEF.

2. Materials and Methods

2.1. Study Population

We prospectively and consecutively enrolled all patients diagnosed with HFrEF undergoing Optimizer Smart® implant between November 2019 and November 2021.

The following inclusion criteria were used:

- HFrEF with a left ventricular ejection fraction < 40%;
- NYHA class II–III;
- QRS duration < 120 ms.
- The following exclusion criteria were used:
- Acute coronary syndrome in the previous three month;
- Non-optimal medical therapy with disease modifier drugs for HFrEF.

Demographic, clinical, and laboratory data were acquired from stable patients 24 h before device implantation.

The research was conducted according to the Declaration of Helsinki and approved by the ethics committee of the AORN dei Colli-Ospedale Monaldi (resolution No. 903/2020).

Signed informed consent was obtained for all patients.

2.2. Optimizer Smart Implant

Implantation of the Optimizer Smart® (Impulse Dynamics Inc., Marlton, NJ, USA) was performed under local anaesthesia.

Two electrodes, required to detect ventricular activity and subsequent delivery of CCM signals, were placed on the right interventricular septum via the subclavian vein.

The ends of both ventricular electrodes were actively attached to the right side of the interventricular septum at least 3 cm away from the implantable defibrillator lead.

Finally, both leads were connected to the Smart Optimizer®, and the device was implanted in a subcutaneous pocket (Figure 1)

2.3. Echocardiographic Evaluation

Standard transthoracic echocardiography and Doppler assessment were performed with available market equipment (Vivid E9—GE Healthcare, Chicago, IL, USA) as recommended elsewhere [12,13].

Two expert echocardiographers analysed all echocardiographic studies, and the value of specific measurements was obtained from the average of 3–5 cardiac cycles.

Systolic excursion in the tricuspid annular plane (TAPSE) was measured in the optimized 4-chamber apical view to obtain the correct M-mode orientation.

Myocardial systolic excursion velocity (S′) was measured at the lateral tricuspid annulus with pulsed tissue Doppler imaging (TDI).

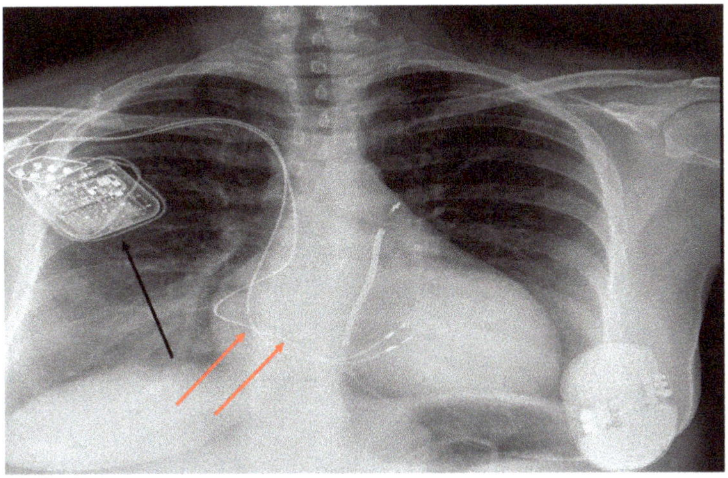

Figure 1. An example of the Optimizer Smart® implant in a female patient with a subcutaneous ICD. Note the presence of the impulse pocket generator (black arrow) and of two electrodes for the delivery of CCM therapy (red arrows).

The pulmonary artery systolic pressure (PASP) was inferred from the maximum velocity of the tricuspid regurgitant jet (in accordance with the modified Bernoulli equation). Adding to the obtained value, the right atrial pressure was calculated according to the size and collapsibility of the inferior vena cava following international recommendations. LV and LA volumes were measured using the modified Simpson's rule with biplane planimetry [13].

Right ventricular myocardial strain parameters were measured using the Q-Analysis software package (EchoPAC BT2.02; GE Vingmed, Horten, Norway). After manually identifying the end-systolic endocardial boundary of the right ventricle by locating three points (two on the tricuspid annulus) and one on the apex of the right ventricle, a region of interest (ROI) was automatically generated.

Next, the width of the ROI was manually adjusted to include the entire myocardial wall. Finally, according to international recommendations, we calculated the RV free-wall longitudinal strain (RVFWLS) value by averaging the values obtained from the three free-wall segments.

Right ventricular arterial coupling was estimated as the ratio of the TAPSE and PASP value.

Echocardiographic evaluations were performed for all patients 24 h before and 6 months after Optimizer Smart® implantation.

2.4. Statistical Analysis

Statistical analyses were performed using Prism 9 (GraphPad Software, San Diego, CA, USA).

Demographic and clinical variables were expressed as medians and standard deviations. Qualitative variables were expressed as both numbers and percentages. Wilcoxon-rank test was used to compare the differences between values at baseline and treatment values in case of non-normal distribution; a *t*-test was used for variables with normal distribution. All *p*-values were two-sided; $p < 0.05$ indicated statistical significance.

ROC (Receiver Operating Feature) curve analysis was performed to select the cut-off of echocardiographic measures. The reproducibility of measurements was determined in all patients, considering inter-observer and intra-observer variability by using intraclass correlation (ICC).

3. Results

During the study period, we enrolled 21 patients with HFrEF who underwent Optimizer Smart implant®. Demographic and clinical characteristics are reported in Table 1.

Table 1. Baseline demographic, clinical, and echocardiographic patients' characteristics.

Variable	Overall Population (21)
Age (mean ± SD)	65 ± 12.5 years
Female sex (n, %)	3 (14.2%)
Ischemic (n, %)	12 (57%)
Hypertension (n, %)	10 (47%)
Diabetes (n, %)	7 (33%)
COPD (n, %)	6 (28 %)
NYHA class III (n, %)	13 (62%)
NYHA class IV (n, %)	8 (38%)
ICD-DR (n, %)	16 (71%)
S-ICD	2 (9%)
CRT-D	3 (14%)
SBP (mean ± SD)	108 ± 18 mmHg
DBP (mean ± SD)	65 ± 9 mmHg
NT-pro BNP (mean ± SD)	2665 ± 1298 pg/mL
Atrial fibrillation	7 (33%)
LVEDV (mean ± SD)	224.2 ± 69.8 mL
LVESV(mean ± SD)	154.8 ± 53.6 mL
LVEF (mean ± SD)	30.2 ± 6.1%
LAVi	47.2 ± 7.9 mL/m^2
Loop diuretic (n, %)	14 (66%)
Beta-Blockers (n, %)	21 (100%)
ARNI (n %)	21 (100%)
MRA (n, %)	16 (76%)

Of the overall population enrolled, 9 patients (42%) had an ischemic form of HFrEF, (left ventricular ejection 29 ± 6%) and 4 patients (19%) had a low TAPSE (<16 mm) and S wave (<10 cm/s) values.

At six months, 2 patients (10%) died, while the follow-up echocardiographic data were collected for 19 patients (90%). No significant changes in disease modifiers drugs doses occurred during follow up (Table 2). The r coefficient value for intra-observed variability was 0.82 and 0.78 for inter-observed variability.

Table 2. Comparison of disease modifiers drugs dosage before and after CCM implants.

Drugs	Baseline Dose (Mean ± SD)	Follow-Up Dose (Mean ± SD)	*p*-Value
Bisoprolol	7.8 ± 1.9 mg	7.3 ± 2.1 mg	0.163
Carvedilol	40.8 ± 9.8 mg	43.3 ± 11.4 mg	0.098
Sacubitril/valsartan	95.4 ± 58.7 mg	103.8 ± 62.5 mg	0.087
Eplerenone	22.8 ± 12.6 mg	20.5 ± 10.2	0.34

3.1. Effects of CCM on RV Reverse Remodeling and Contractility Index

The echocardiographic index of RV contractility improved in patients enrolled in the study during follow up.

There was a significant reduction in RV end-diastolic transversal diameters ($p < 0.05$) (Table 3), while the severity of tricuspid regurgitation remained stable (Table 3).

Table 3. Effects of CCM on LV and RV echocardiographic variables at six months follow up.

Variable	Baseline	6 Months Follow Up	*p*-Value
LVEDV (mL)	224.2 ± 69.8	198.3 ± 45.7	<0.05
LVESV (mL)	154.8 ± 53.6	122.6 ± 66.3	<0.05
LVEF (%)	30.2 ± 6.1	35.4 ± 7.3	<0.05
TAPSE (mm)	16.6 ± 4.2	18.5 ± 3.6	<0.05
S wave (cm/s)	10.1 ± 1.8	11.3 ± 1.4	<0.05
PASP (mmHg)	34.2 ± 9.6	28.1 ± 6.9	<0.05
RVFWLS (%)	−13.7 ± 2.5	−15.1 ± 2.8	<0.05
TAPSE/PASP (mm/mmHg)	0.52 ± 0.22	0.66 ± 0.21	<0.05
RVOT PLAX (mm)	28.2 ± 3.1	27.1 ± 4.2	0.062
RVD 1	26.8 ± 5.3	25.7 ± 4.1	<0.05
RVD 2	28.1 ± 4.3	26.2 ± 3.2	<0.05
TI mild (n, %)	16 (76%)	18 (85%)	NA
TI moderate (n, %)	3 (14%)	2 (9%)	NA
TI severe (n, %)	2 (10%)	1 (4%)	NA

As shown in Figures 2 and 3, a statistically significant increase in both TAPSE (16.6 ± 4.2 mm vs. 18.5 ± 3.6 mm; $p < 0.05$) and S wave (10.1 ± 1.8 cm/s vs. 11.3 ± 11.4 cm/s; $p < 0.05$) occurred at six months.

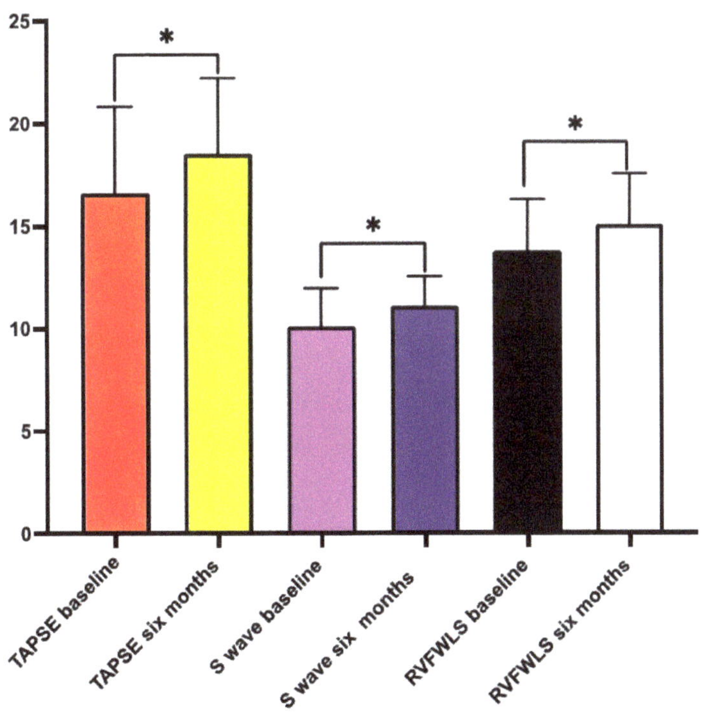

Figure 2. Effects of CCM therapy of right ventricular contractility indexes. * = $p < 0.05$.

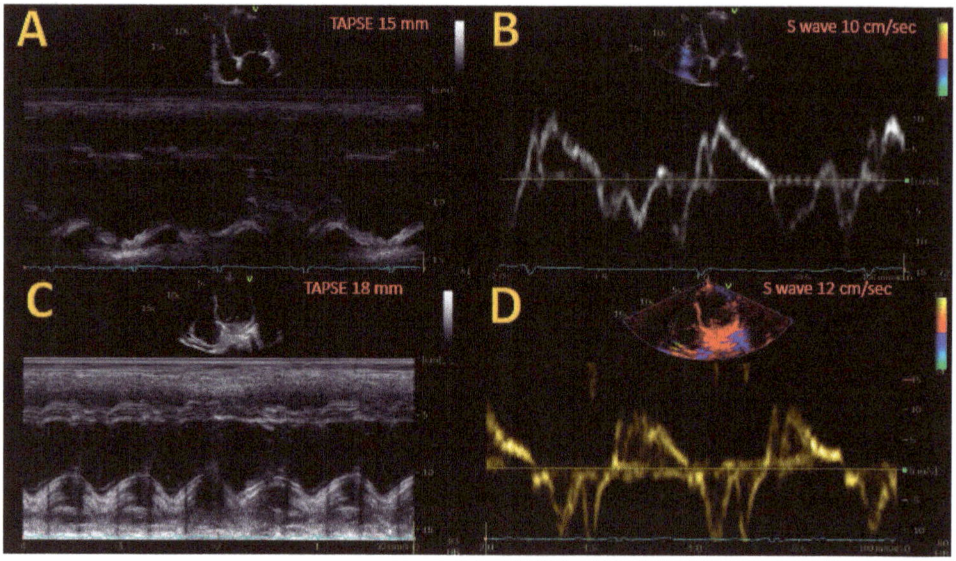

Figure 3. Example of the effects of CCM on TAPSE (panel (**A**,**C**)) and S wave (panel (**B**,**D**)) at six months follow-up.

Concurrently, a significant increase in more specific and reproducible index of RV function, the RVFWLS occurred (−13.7 ± 1.8% vs. −15.6 ± 2.3%; $p < 0.05$; Figure 3).

Notably, the increase in all the parameters of RV function occurred in patients with a standard RV function rather than in patients with reduced ventricular function at baseline ($p = 0.12$).

3.2. Effects of CCM on PASP

Positive effects of CCM on right ventricular performance determine hemodynamic benefits, too.

As shown in Figure 4, CCM determines, at six months, a significant reduction in PASP (34.2 ± 9.8 mmHg vs. 28 ± 6.2 mmHg; $p < 0.05$); also, in this case, these effects are present both in patients with normal values at baseline and in patients with a high PASP value at enrollment ($p = 0.78$).

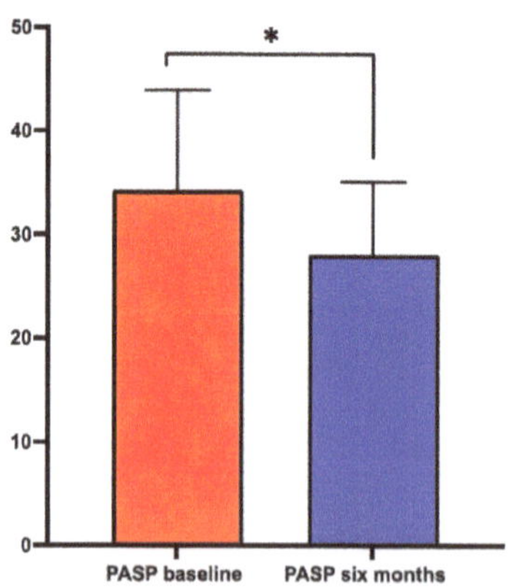

Figure 4. Effects of CCM on pulmonary artery systolic pressure. * = $p < 0.05$.

3.3. *Effects of CCM on RV-PA Coupling*

The positive effects that CCM therapy brings about both in terms of increasing right ventricular contractility indices and reducing pressures in the pulmonary circulation result in improved coupling between RV and PA. In fact, at six months of follow up, the TAPSE/PAPS ratio improves significantly (0.52 ± 0.14 mm/mmHg vs. 0.66 ± 0.23 mm/mmHg; $p < 0.05$; Figure 5).

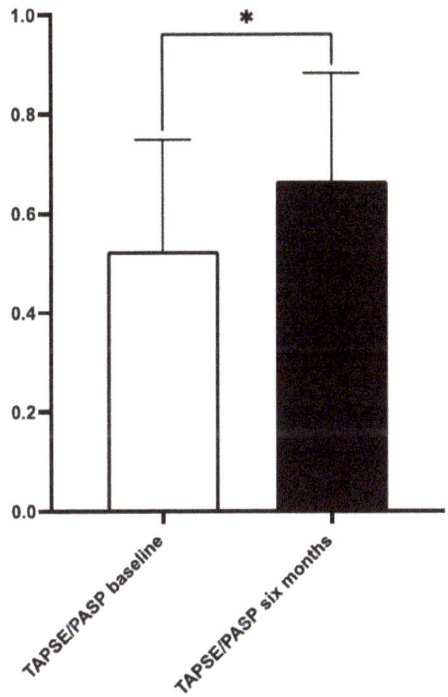

Figure 5. Effects of CCM on right ventricular arterial coupling. * = $p < 0.05$.

4. Discussion

For the first time, our data show that in patients with HFrEF, CCM therapy can improve after six months of follow-up RV systolic function, as evidenced by an increase in both tapes, RVS wave value and RVFWLS, which resulted in RV reverse remodeling. All three indices of RV function increased in both normal and reduced RV function patients. At six months follow up, the degree of tricuspid regurgitation remained stable.

At six months, CCM also determined a significant reduction in PASP in patients with normal values at baseline and those with high PASP values at enrollment.

Furthermore, the positive effects of CCM in terms of increasing RV contractility and reducing pulmonary pressures resulted in improved coupling between RV and PA, as evidenced by an increase in the TAPSE/PAPS ratio.

4.1. *Improvement of RV Reverse Remodeling Systolic Function after CCM*

Although several studies reported improvements in symptoms, exercise capacity, and LV global and regional LV contractility, whether CCM would induce favorable changes in RV systolic function was largely unknown.

As guidelines [13] recommended, RV size was measured from a four-chamber view in the context of the left ventricle. Both basal and mid-cavity diameters were reduced, suggesting a favorable RV remodeling; however, the severity of tricuspid regurgitation remained stable, maybe because of the presence of leads passing from the right atrium to the RV through the valve.

In this study, we found that all three RV systolic function indices (TAPSE, S wave, RVFWLS) improved after 6 months of CCM therapy.

TAPSE is a measure of longitudinal contraction of the RV. It requires no geometric assumptions, it has the advantage of being obtained even with poor images, and it can be measured on all ultrasound machines [14].

TAPSE has prognostic value in a variety of conditions that may affect the RV, such as HFrEF [15] and pulmonary hypertension [16,17].

TDI-derived lateral tricuspid annulus S-wave velocity is easy to measure, reliable, and reproducible. Like TAPSE, it is a unidimensional measure and evaluates RV longitudinal function [16].

Unlike wall motion displacement (TAPSE) or velocity (S wave), RV free-wall myocardial strain (myocardial deformation) is unaffected by the motion of the entire heart and allows distinction between active and passive myocardial tissue movement. RVFWLS is an angle-independent speckle-tracking echocardiography (STE)-derived parameter that is very useful for the evaluation of RV contractile function. It has prognostic value for cardiovascular morbidity and mortality [18,19].

Therefore, as previously demonstrated for LV global and regional systolic function [20], in our study the CCM also seems to improve RV systolic function, increasing global RV contractility, not just limited to the septal wall where CCM signals were delivered.

Similar to the findings in our study, in an animal model of chronic right ventricular dysfunction, CCM has been shown to improve global cardiac function significantly inboth ventricles [21].

CCM seems to be useful in HFrEF patients to improve both LV and RV systolic function, probably given its proven effects on regional and global myocardial contractility [20] and given the key role played by the interventricular septum in the contractility of both ventricles via ventricular interdependence [15,22,23]. Furthermore, Yucel G et al. showed better positive biventricular echocardiographic results in patients with lower LVEF, probably because the intensity of LVEF reduction appears to be correlated partly with impaired neuro-humoral activation associated with HF, assuming that the underlying mechanism could be "the worse the LVEF at baseline, the more can be repaired" [24].

4.2. *Improvement of PASP after CCM*

Similar to a recent case report, which shows the positive effect of CCM on pulmonary pressure measured by Cardio MEMS [25], in this study we found that CCM determined a significant reduction in PASP at six months. The favorable effect of CCM on pulmonary pressure would be due to the improvement in both ventricles' contractility.

4.3. *Improvement of RV-PA Coupling after CCM*

We found an improvement RV-PA coupling after 6 months of CCM, evidenced by the increase TAPSE/PASP ratio, which incorporates both RV longitudinal displacement and load. This index assesses RV contraction by plotting fiber longitudinal shortening versus the force generated to overcome the imposed load [26,27].

Unlike TAPSE, S wave and FWRVLS depend on RV loading conditions. TAPSE/PASP ratio allows estimation of RV performance, and it is also a non-invasive index of RV to pulmonary circulation coupling based on the correlation with invasively evaluated RV systolic elastance/arterial elastance [27]. This echocardiographic index is a predictor of mortality in patients with HFrEF and with severe PAH [28,29].

5. Study Limitations

The relatively small sample size, single-centre study design, and observational nature of the study may affect our results. STE can be affected by RV loading conditions.

We do not assess RV with 3D echocardiography.

6. Conclusions

At six months follow up, CCM therapy increased RV performance, improving RV systolic function, PASP, and coupling between RV and PA. A better forward ejection of blood could be useful for RV reverse remodeling.

Additional larger studies are needed to provide a greater understanding of the long-term impact of CCM on RV.

Author Contributions: Conceptualization, C.C. and S.D.V.; and methodology, C.C., S.D.V., M.L.M., A.D., E.A. and G.N.; data curation, C.C., M.L.M., V.E. and D.M.; writing—original draft preparation, C.C., M.L.M. and D.M.; writing—review and editing, C.C., S.D.V., A.D., E.A., G.N. and G.P.; visualization, C.C.; supervision, G.P. and D.M. All authors have read and agreed to the published version of the manuscript.

Funding: This research received no external funding.

Institutional Review Board Statement: The study was conducted in accordance with the Declaration of Helsinki, andapproved byethics committee of the AORN dei Colli-Ospedale Monaldi (resolution No. 903/2020).

Informed Consent Statement: Informed consent was obtained from all subjects involved in the study.

Data Availability Statement: The data presented in this study are available on request from the corresponding author.

Conflicts of Interest: The authors declare no conflict of interest.

References

1. Savarese, G.; Lund, L.H. Global Public Health Burden of Heart Failure. *Card. Fail. Rev.* **2017**, *3*, 7–11. [CrossRef]
2. Benjamin, E.J.; Virani, S.S.; Callaway, C.W.; Chamberlain, A.M.; Chang, A.R.; Cheng, S.; Chiuve, S.E.; Cushman, M.; Delling, F.N.; Deo, R.; et al. Heart Disease and Stroke Statistics-2018 Update: A Report from the American Heart Association. *Circulation* **2018**, *137*, e67–e492. Erratum in *Circulation* **2018**, *137*, e493. [CrossRef]
3. Komajda, M.; Böhm, M.; Borer, J.S.; Ford, I.; Tavazzi, L.; Pannaux, M.; Swedberg, K. Incremental benefit of drug therapies for chronic heart failure with reduced ejection fraction: A network meta-analysis. *Eur. J. Heart Fail.* **2018**, *20*, 1315–1322. [CrossRef] [PubMed]
4. Moskovitch, J.; Voskoboinik, A. Cardiac resynchronization therapy: A comprehensive review. *Minerva Medica* **2019**, *110*, 121–138. [CrossRef] [PubMed]
5. Abraham, W.T. Cardiac Resynchronization Therapy and Cardiac Contractility Modulation in Patients with Advanced Heart Failure. *Heart Fail. Clin.* **2021**, *17*, 599–606. [CrossRef]
6. Campbell, C.M.; Kahwash, R.; Abraham, W.T. Optimizer Smart in the treatment of moderate-to-severe chronic heart failure. *Future Cardiol.* **2020**, *16*, 13–25. [CrossRef]
7. Lawo, T.; Borggrefe, M.; Butter, C.; Hindricks, G.; Schmidinger, H.; Mika, Y.; Burkhoff, D.; Pappone, C.; Sabbah, H.N. Electrical Signals Applied during the Absolute Refractory Period: An Investigational Treatment for Advanced Heart Failure in Patients with Normal QRS Duration. *J. Am. Coll. Cardiol.* **2005**, *46*, 2229–2236. [CrossRef]
8. Butter, C.; Rastogi, S.; Minden, H.-H.; Meyhöfer, J.; Burkhoff, D.; Sabbah, H.N. Cardiac Contractility Modulation Electrical Signals Improve Myocardial Gene Expression in Patients with Heart Failure. *J. Am. Coll. Cardiol.* **2008**, *51*, 1784–1789. [CrossRef]
9. Zhang, F.; Dang, Y.; Li, Y.; Hao, Q.; Li, R.; Qi, X. Cardiac Contractility Modulation Attenuate Myocardial Fibrosis by Inhibiting TGF-β1/Smad3 Signaling Pathway in a Rabbit Model of Chronic Heart Failure. *Cell. Physiol. Biochem.* **2016**, *39*, 294–302. [CrossRef]
10. Lyon, A.R.; Samara, M.A.; Feldman, D.S. Cardiac contractility modulation therapy in advanced systolic heart failure. *Nat. Rev. Cardiol.* **2013**, *10*, 584–598. Erratum in *Nat. Rev. Cardiol.* **2014**, *11*, 188. [CrossRef]
11. Cappannoli, L.; Scacciavillani, R.; Rocco, E.; Perna, F.; Narducci, M.L.; Vaccarella, M.; D'Amario, D.; Pelargonio, G.; Massetti, M.; Crea, F.; et al. Cardiac contractility modulation for patient with refractory heart failure: An updated evidence-based review. *Heart Fail. Rev.* **2021**, *26*, 227–235. [CrossRef] [PubMed]
12. Nagueh, S.F.; Smiseth, O.A.; Appleton, C.P.; Byrd, B.F., 3rd; Dokainish, H.; Edvardsen, T.; Flachskampf, F.A.; Gillebert, T.C.; Klein, A.L.; Lancellotti, P.; et al. Recommendations for the Evaluation of Left Ventricular Diastolic Function by Echocardiography: An Update from the American Society of Echocardiography and the European Association of Cardiovascular Imaging. *Eur. Heart J. Cardiovasc. Imaging* **2016**, *17*, 1321–1360. [CrossRef] [PubMed]
13. Lang, R.M.; Badano, L.P.; Mor-Avi, V.; Afilalo, J.; Armstrong, A.; Ernande, L.; Flachskampf, F.A.; Foster, E.; Goldstein, S.A.; Kuznetsova, T.; et al. Recommendations for Cardiac Chamber Quantification by Echocardiography in Adults: An Update from the American Society of Echocardiography and the European Association of Cardiovascular Imaging. *Eur. Heart J. Cardiovasc. Imaging* **2015**, *16*, 233–271. Erratum in *Eur. Heart J. Cardiovasc. Imaging* **2016**, *17*, 412. [CrossRef] [PubMed]

14. Jones, N.; Burns, A.T.; Prior, D.L. Echocardiographic Assessment of the Right Ventricle–State of the Art. *Heart Lung Circ.* **2019**, *28*, 1339–1350. [CrossRef]
15. Santamore, W.P.; Dell'Italia, L.J. Ventricular interdependence: Significant left ventricular contributions to right ventricular systolic function. *Prog. Cardiovasc. Dis.* **1998**, *40*, 289–308. [CrossRef]
16. Rudski, L.G.; Lai, W.W.; Afilalo, J.; Hua, L.; Handschumacher, M.D.; Chandrasekaran, K.; Solomon, S.D.; Louie, E.K.; Schiller, N.B. Guidelines for the Echocardiographic Assessment of the Right Heart in Adults: A Report from the American Society of Echocardiography: Endorsed by the European Association of Echocardiography, a registered branch of the European Society of Cardiology, and the Canadian Society of Echocardiography. *J. Am. Soc. Echocardiogr.* **2010**, *23*, 685–713. [CrossRef]
17. Kind, T.; Mauritz, G.-J.; Marcus, J.T.; van de Veerdonk, M.; Westerhof, N.; Vonk-Noordegraaf, A. Right ventricular ejection fraction is better reflected by transverse rather than longitudinal wall motion in pulmonary hypertension. *J. Cardiovasc. Magn. Reson.* **2010**, *12*, 35. [CrossRef]
18. Dandel, M.; Hetzer, R. Echocardiographic strain and strain rate imaging–Clinical applications. *Int. J. Cardiol.* **2009**, *132*, 11–24. [CrossRef]
19. Badano, L.P.; Kolias, T.J.; Muraru, D.; Abraham, T.P.; Aurigemma, G.; Edvardsen, T.; D'Hooge, J.; Donal, E.; Fraser, A.G.; Marwick, T.; et al. Standardization of left atrial, right ventricular, and right atrial deformation imaging using two-dimensional speckle tracking echocardiography: A consensus document of the EACVI/ASE/Industry Task Force to standardize deformation imaging. *Eur. Heart J. Cardiovasc. Imaging* **2018**, *19*, 591–600. Erratum in *Eur. Heart J. Cardiovasc. Imaging* **2018**, *19*, 830–833. [CrossRef]
20. Yu, C.-M.; Chan, J.Y.-S.; Zhang, Q.; Yip, G.W.; Lam, Y.-Y.; Chan, A.; Burkhoff, D.; Lee, A.; Fung, J.W.-H. Impact of Cardiac Contractility Modulation on Left Ventricular Global and Regional Function and Remodeling. *JACC Cardiovasc. Imaging* **2009**, *2*, 1341–1349. [CrossRef]
21. Roubertie, F.; Eschalier, R.; Zemmoura, A.; Thambo, J.-B.; Rooryck, C.; Labrousse, L.; Ploux, S.; Ritter, P.; Haïssaguerre, M.; Dos Santos, P.; et al. Cardiac Contractility Modulation in a Model of Repaired Tetralogy of Fallot: A Sheep Model. *Pediatr. Cardiol.* **2016**, *37*, 826–833. [CrossRef] [PubMed]
22. Sanz, J.; Sánchez-Quintana, D.; Bossone, E.; Bogaard, H.J.; Naeije, R. Anatomy, Function, and Dysfunction of the Right Ventricle. *J. Am. Coll. Cardiol.* **2019**, *73*, 1463–1482. [CrossRef] [PubMed]
23. Haddad, F.; Hunt, S.A.; Rosenthal, D.; Murphy, D.J. Right Ventricular Function in Cardiovascular Disease, Part I. *Circulation* **2008**, *117*, 1436–1448. [CrossRef] [PubMed]
24. Yücel, G.; Fastner, C.; Hetjens, S.; Toepel, M.; Schmiel, G.; Yazdani, B.; Husain-Syed, F.; Liebe, V.; Rudic, B.; Akin, I.; et al. Impact of baseline left ventricular ejection fraction on long-term outcomes in cardiac contractility modulation therapy. *Pacing Clin. Electrophysiol.* **2022**, *45*, 639–648. [CrossRef] [PubMed]
25. Visco, V.; Esposito, C.; Manzo, M.; Fiorentino, A.; Galasso, G.; Vecchione, C.; Ciccarelli, M. A Multistep Approach to Deal with Advanced Heart Failure: A Case Report on the Positive Effect of Cardiac Contractility Modulation Therapy on Pulmonary Pressure Measured by CardioMEMS. *Front. Cardiovasc. Med.* **2022**, *9*, 874433. [CrossRef]
26. Guazzi, M.; Bandera, F.; Pelissero, G.; Castelvecchio, S.; Menicanti, L.; Ghio, S.; Temporelli, P.L.; Arena, R. Tricuspid annular plane systolic excursion and pulmonary arterial systolic pressure relationship in heart failure: An index of right ventricular contractile function and prognosis. *Am. J. Physiol. Circ. Physiol.* **2013**, *305*, H1373–H1381. [CrossRef]
27. Guazzi, M.; Dixon, D.; Labate, V.; Beussink-Nelson, L.; Bandera, F.; Cuttica, M.J.; Shah, S.J. RV Contractile Function and its Coupling to Pulmonary Circulation in Heart Failure with Preserved Ejection Fraction. *JACC Cardiovasc. Imaging* **2017**, *10*, 1211–1221. [CrossRef]
28. Amsallem, M.; Boulate, D.; Aymami, M.; Guihaire, J.; Selej, M.; Huo, J.; Denault, A.Y.; McConnell, M.V.; Schnittger, I.; Fadel, E.; et al. Load Adaptability in Patients with Pulmonary Arterial Hypertension. *Am. J. Cardiol.* **2017**, *120*, 874–882. [CrossRef]
29. Topilsky, Y.; Oh, J.K.; Shah, D.K.; Boilson, B.A.; Schirger, J.A.; Kushwaha, S.S.; Pereira, N.L.; Park, S.J. Echocardiographic Predictors of Adverse Outcomes After Continuous Left Ventricular Assist Device Implantation. *JACC Cardiovasc. Imaging* **2011**, *4*, 211–222. [CrossRef]

applied sciences

MDPI

Article

New Possibilities in Heart Failure: The Effects of Tadalafil on Diastolic Function in Patients Undergoing Robot-Assisted Radical Prostatectomy

Felice Crocetto [1], Vittoria Cuomo [2], Ciro Santoro [3,*], Ludovica Fiorillo [2], Biagio Barone [1], Davide Arcaniolo [4], Teresa Fedele [2], Ciro Imbimbo [1] and Roberta Esposito [2]

1 Department of Neurosciences, Reproductive Sciences and Odontostomatology, Federico II University Hospital, 80131 Naples, Italy; felice.crocetto@unina.it (F.C.); biagio.barone@unina.it (B.B.); ciro.imbimbo@unina.it (C.I.)
2 Department of Clinical Medicine and Surgery, Federico II University Hospital, 80131 Naples, Italy; vittycuomo@yahoo.it (V.C.); ludovica.fiorillo91@gmail.com (L.F.); teresafedele88@libero.it (T.F.); roberta.esposito1@unina.it (R.E.)
3 Department of Advanced Biomedical Sciences, Federico II University Hospital, 80131 Naples, Italy
4 Urology Unit, Department of Woman, Child and General and Specialized Surgery, University of Campania "Luigi Vanvitelli", 80131 Naples, Italy; davide.arcaniolo@gmail.com
* Correspondence: ciro.santoro@unina.it; Tel.: +39-0817463663

Abstract: Inhibitors of phosphodiesterase type 5 (PDE5i) are the first-line treatment for erectile dysfunction and are also used to treat pulmonary hypertension. PDE5i impedes the breakdown of nitric oxide (NO)-driven cyclic guanosine monophosphate (cGMP) in smooth muscle cells of the vascular bed, acting as a potent vasodilator. In heart failure, cGMP signaling is altered. The modulation of cGMP has therefore emerged as a potential therapeutic option for heart failure. In this prospective observational study, we aim to investigate whether tadalafil, a long-acting PDE5i used for erectile dysfunction, could also improve diastolic function assessed by cardiac ultrasound. A total of 23 patients were enrolled, undergoing nerve-sparing robot-assisted radical prostatectomy for prostate cancer and treated with 20 mg tadalafil on alternate days to recover erectile function. All patients underwent tadalafil treatment for at least 6 months. Participants underwent a clinical and cardiac ultrasound with color Doppler assessment at baseline, after 3 months, and after 6 months. At 6 months, no significant difference was found apart from lower E/e′ ratio (7.4 ± 2.7 vs. 6.3 ± 1.3; $p < 0.03$), peak velocity of TR jet (2.4 ± 0.2 vs. 2.1 ± 0.2; $p < 0.001$), and PAPs (27.3 ± 3.6 vs. 22.9 ± 5.7; $p < 0.005$). Our prospective study shows that 6 months of erectile dysfunction therapy for secondary to radical prostatectomy is associated with a favorable effect on diastolic function, improving the E/e′ ratio and peak velocity of the TR jet.

Keywords: diastolic disfunction; echocardiography; erectile dysfunction; nitric oxide signaling

Citation: Crocetto, F.; Cuomo, V.; Santoro, C.; Fiorillo, L.; Barone, B.; Arcaniolo, D.; Fedele, T.; Imbimbo, C.; Esposito, R. New Possibilities in Heart Failure: The Effects of Tadalafil on Diastolic Function in Patients Undergoing Robot-Assisted Radical Prostatectomy. *Appl. Sci.* **2022**, *12*, 5629. https://doi.org/10.3390/app12115629

Academic Editor: Julio Garcia Flores

Received: 6 April 2022
Accepted: 30 May 2022
Published: 1 June 2022

1. Introduction

Inhibitors of phosphodiesterase type 5 (PDE5is) are a class of drugs that suppresses the degradation of cyclic guanosine monophosphate (cGMP) and were initially studied for the treatment of angina pectoris [1] before being considered and employed as first-line therapy for erectile dysfunction and pulmonary hypertension. The most commonly used PDE5is are sildenafil, vardenafil, tadalafil, and avanafil [2]. The enhancement of cGMP, derived from nitric oxide, leads to the distension of vascular smooth muscle and vasodilatation, which justifies the therapeutic efficacy in erectile dysfunction and pulmonary hypertension [3]. In addition to the therapeutic role of PDE5is in erectile dysfunction and pulmonary hypertension, preclinical studies showed that PDE5i could likely play a role in mediating cardioprotection through complex mechanisms, suggesting a possible clinical role of PDE5i for the treatment of cardiovascular diseases [4–8]. Patients affected by prostate cancer very frequently need cardiovascular assessment based on the increased risk of cardiovascular side effects during hormonal treatment [9]. Cross links between urologic treatment and cardiovascular disease are well recognized. It is indeed well established that cGMP signaling is abnormal in heart failure [10]; therefore, the modulation of cGMP has emerged as a potential therapeutic strategy for heart failure. However, no clear evidence supports PDE5i therapy for heart failure due to the controversial results obtained by trials investigating the role of a cGMP-augmenting therapy with PDE5i in the treatment of heart failure [8,10]. Due to these premises, further research is urgently needed to clarify the role of cGMP-augmenting therapies across the entire spectrum of heart failure, possibly even in the preclinical phase.

In the current prospective study, we aimed to investigate whether tadalafil, a long-acting PDE5i, improves the diastolic function. This was assessed using cardiac ultrasound in patients with no cardiovascular diseases, taking PDE5i for the prevention of erectile dysfunction after nerve-sparing robot-assisted radical prostatectomy (RARP) for prostate cancer.

2. Materials and Methods

2.1. Study Population

We conducted a prospective observational study involving patients who underwent nerve-sparing robot-assisted radical prostatectomy for prostate cancer at Federico II University Hospital, Naples, Italy (Research Ethics Board of "Federico II" University of Naples, n.316/20), from November 2018 to December 2019. Patients in treatment with tadalafil for erectile function recovery, i.e., penile rehabilitation, for at least 6 months after prostatectomy were included in the study [11]. Exclusion criteria were considered: diabetes mellitus, left ventricular (LV) ejection fraction <53%, more than mild valvular disease, atrial fibrillation, history of dyspnea, and inadequate echocardiographic imaging. All the patients underwent thorough cardiac clinical assessment before starting tadalafil therapy and were assessed again at 3- and 6-month follow-up visits. These visits included medical history, clinical examination, heart rate measurement, blood pressure, body mass index, and color Doppler cardiac ultrasound, performed at the Interdepartmental Laboratory of Echocardiography at the same University. All patients were treated with tadalafil 20 mg every other day for at least 6 months. All patients provided written informed consent.

2.2. Transthoracic Echocardiography

An experienced echocardiologist (L.F.) performed all image acquisitions using a GE Vivid E95 ultrasound machine (GE Healthcare, Horten, Norway) provided by a 3.5 MHz transducer. LV quantitative analysis was conducted in accordance with guidelines [12]. LV mass and relative wall thickness were calculated using a two-dimensional parasternal long-axis view. LV mass was indexed to the patient's height elevated to 2.7, while the left atrial (LA) volume was indexed to the body surface area. LV ejection fraction (LVEF) was determined using the biplane method computing the LV end-diastolic and end-systolic volumes in apical four- and two-chamber views. LV diastolic function was measured according to current recommendations [13]. Global Longitudinal Strain (GLS) was assessed offline using a workstation (EchoPAC only software version 113, GE Healthcare, Horten, Norway), with images taken in the apical views (three-chamber, four-chamber, and two-chamber). An automated 2D strain software with manual adjustment capability was used to trace endocardial and epicardial borders. Peak longitudinal strain was measured for the basal, mid, and apical segments for each wall, obtaining a 16-segment bullseye. GLS was obtained as a mean of all peak strain before the aortic valve closure. The reproducibility of GLS was previously published [14,15]. The day-to-day variability in the measurements of parameters investigating LV diastolic function was previously assessed in our laboratory by calculating intraclass correlation coefficients [16]. Our results indicated excellent reliability of the estimates of LV diastolic function.

RV Systolic Function

Tricuspid annular plane systolic excursion (TAPSE) was measured to assess RV systolic function. Pulmonary arterial systolic pressure (PASP) was estimated in accordance with current guidelines; tricuspid regurgitation peak velocity was added to an estimate of right atrial pressure (RAP). RAP was estimated from the measurement of the size and the respiratory reactivity of the inferior vena cava (IVC): (a) RAP = 5 mmHg when IVC diameter is <2.1 cm and has normal inspiratory collapse; (b) RAP = 10 mmHg when IVC diameter is >2.1 cm or when it collapses <50%; (c) RAP = 15 mmHg when IVC is both dilated and collapses <50%; (d) RAP = 20 mmHg when IVC is dilated and has no visible collapse [17].

2.3. Statistical Analysis

Continuous variables are expressed as mean ± standard deviation. Statistical differences between continuous variables were tested using the unpaired Student's *t*-test or the equivalent nonparametric procedure (Mann–Whitney U test) for variables not normally distributed. A comparison of echocardiographic variables before and after the use of tadalafil was performed using the paired *t*-test. A *p*-value of <0.05 was considered statistically significant. The day-to-day variability in chosen variables was assessed by calculating intraclass correlation coefficients (ICCs) and their 95% CIs. Statistical analysis was performed using SPSS package, release 12 (SPSS Inc., Chicago, IL, USA).

3. Results

Of a total of 49 subjects, we excluded 19 because of: LVEF < 53% (n = 4), more than mild valvular disease (n = 4), atrial fibrillation (n = 2), inadequate echocardiographic imaging (n = 3), and drop-out at follow-up (n = 10). The remaining 30 patients were included in this analysis (Figure 1).

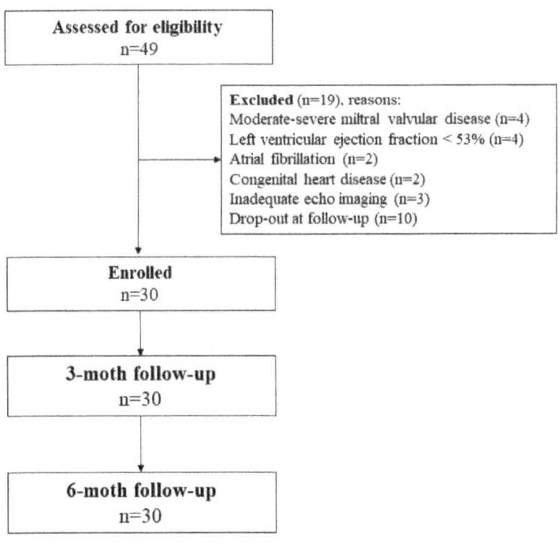

Figure 1. Patient flow and study design.

The mean age was 64 + 7.6 years. At baseline, 11 patients had arterial hypertension, 7 were dyslipidemic, 8 were smokers, none were diagnosed with chronic obstructive pulmonary disease (COPD) and diabetes mellitus, and only 1 patient presented with peripheral arterial disease. Patient characteristics at enrollment and after 6 months of treatment with tadalafil are reported in Table 1.

Table 1. Characteristics of the patients at baseline and after 6 months of treatment with tadalafil.

	Baseline	At 6 Months	p-Value
SBP (mm Hg)	127 ± 13	129 ± 16	0.71
DBP (mm Hg)	73.6 ± 7.9	75.7 ± 8.7	0.33
HR (bpm)	72.0 ± 10.7	69.2 ± 10.5	0.45
BW (kilograms)	78.6 ± 8.5	77.3 ± 9.7	0.49

BW = body weight; DBP = diastolic blood pressure; SBP = systolic blood pressure; HR = heart rate.

Echocardiographic data were collected at 3 and 6 months for all patients. After 3 months of treatment with tadalafil, no difference was found between echocardiographic parameters, as reported in Table 2.

Table 2. Echocardiographic findings at 3 months of treatment with tadalafil.

	Baseline	At 3 Months	p-Value
LVM/ht$^{2.7}$ (g/m$^{2.7}$)	35.0 ± 7.0	36.1 ± 10.1	0.54
RWT	0.37 ± 0.6	0.35 ± 0.5	0.53
E/e' ratio	7.4 ± 2.7	6.9 ± 1.4	0.2
TAPSE (mm)	23.9 ± 3.5	23.2 ± 4.0	0.56
TR vel (m/s)	2.4 ± 0.2	2.3 ± 0.2	0.8
LAVi (mL/m^2)	28.4 ± 6.8	37.8 ± 5.2	0.3
EF (%)	60.4 ±4.9	60.7 ± 3.3	0.9
GLS (%)	21.0 ± 1.6	20.8 ± 1.6	0.7
PAPs (mmHg)	27.3 ± 3.6	26.0 ± 6.0	0.36

LVM/ht$^{2.7}$ = LV mass was indexed to patients' height elevated to 2.7; RWT = relative wall thickness; TAPSE = tricuspid annulus pulsed systolic excursion; TR vel = tricuspid jet velocity; LAVi = left atrial volume index; FE = ejection fraction, GLS = global longitudinal strain; PAPs = pulmonary arterial peak systolic pressure.

At 6 months, there was no significant variation in standard echocardiographic parameters, but E/e' ratio, peak velocity of TR jet, and PAPs were significantly lower, as reported in Table 3 (Figure 2).

Table 3. Echocardiographic findings at 6 months of treatment with tadalafil.

	Baseline	At 6 Months	p-Value
LVM/height$^{2.7}$(g/m$^{2.7}$)	35.0 ± 7.0	37.4 ± 11	0.27
RWT	0.37 ± 0.6	0.36 ± 0.6	0.62
E/e'ratio	7.4 ± 2.7	6.3 ± 1.3	*0.03*
TAPSE (mm)	23.9 ± 3.5	23.0 ± 3.0	0.34
TR vel (m/s)	2.4 ± 0.2	2.1 ± 0.2	*0.001*
LAVi (mL/m^2)	28.4 ± 6.8	27.3 ± 5.6	0.51
EF (%)	60.4 ±4.9	61.0 ± 3.3	0.60
GLS (%)	21.0 ± 1.6	20.8 ± 1.5	0.44
PAPs (mmHg)	27.3 ± 3.6	22.9 ± 5.7	*0.005*

See Table 2 for the legend.

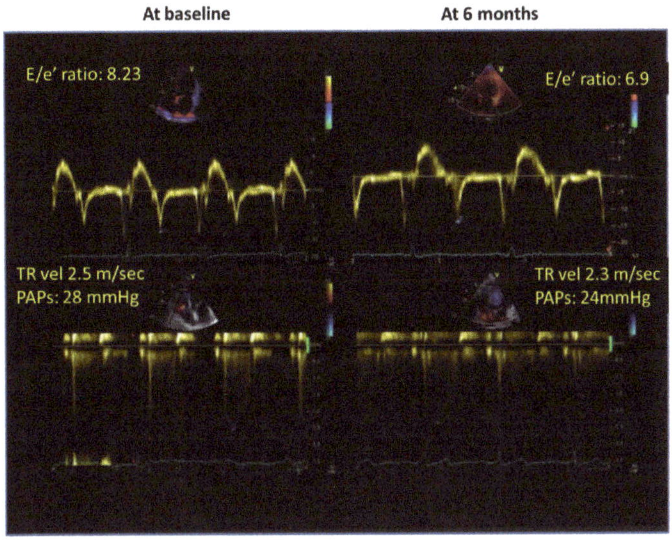

Figure 2. Clinical case of patient undergoing 6 months of tadalafil treatment. Patient's diastolic function at baseline (**left**) panel and after 6 months of treatment (**right**) panel; from top to bottom: pulsed Doppler velocity on mitral valve, tissue Doppler velocity of the lateral mitral annulus, peak systolic tricuspid jet velocity.

4. Discussion

Our prospective study shows that PDE5 inhibition with tadalafil for 6 months, in patients who underwent penile rehabilitation for erectile function secondary to radical prostatectomy, is associated with favorable effects on left ventricular diastolic function, improving the E/e′ ratio and the tricuspid jet (TR). Although over 20 parameters have been suggested to noninvasively assess left ventricular diastolic function, not all these parameters are currently utilized in routine clinical practice [13]. In 2016, the American Society of Echocardiography (ASE) and the European Association of Cardiovascular Imaging (EACVI) proposed a simple algorithm for the noninvasive evaluation of left ventricular diastolic function based on the evaluation of the E/e′ ratio and peak TR velocity [13]. The ESC 2021 guidelines on heart failure confirmed that both the E/e′ ratio and peak TR velocity should be used as echocardiographic markers of left ventricular diastolic dysfunction/raised left ventricular filling pressures [18]. It remains to be proven that the improvement in E/e′ ratio and peak TR velocity observed in our patients after tadalafil therapy may have clinical importance. Indeed, the patients enrolled in our study showed no signs of left ventricular diastolic dysfunction. However, the finding that a drug can improve left ventricular diastolic function may be of interest, although additional investigations dedicated to this issue are needed. Pre-emptive intervention with drugs capable of improving left ventricular diastolic function could prevent the onset of left ventricular diastolic dysfunction in subjects at greater risk or attenuate its progression to heart failure. Data on the beneficial effect of drugs acting on the guanylate cyclase stimulator on patients with chronic HF with reduced ejection fraction have recently been published. Increased levels of cyclic GMP caused by inducing nitric oxide soluble guanylate cyclase (cGM) pathway appear to have a positive impact on hospitalization for heart failure and mortality [19]. Even if no data on nitric oxide pathways induction in patients with HF with preserved systolic function have been provided, a beneficial analogous effect could be supposed for both cGM and PDE drugs that act by increasing nitric oxide levels with a beneficial impact on cardiovascular function. It is known that asymptomatic left ventricular diastolic dysfunction can evolve into overt heart failure. In a randomly selected cohort of

2042 subjects ≥45 years old that were followed for 4 years, Kane et al. demonstrated that diastolic dysfunction was an independent risk factor for heart failure in the elderly, even in apparently healthy subjects [20]. A subsequent meta-analysis of 13 reports based on 11 distinct studies assessing a total of 25,369 participants followed for 7.9 years on average, including asymptomatic subjects with diastolic dysfunction, revealed a 70% increased risk of progressing to heart failure in subjects with asymptomatic diastolic dysfunction in comparison with subjects without asymptomatic diastolic dysfunction [21]. To date, no treatment has been shown to convincingly reduce mortality and morbidity in a population of patients with heart failure with preserved ejection fraction [21]. Therefore, the development of new therapies remains a challenging and unresolved task. The observed improvements in diastolic function indices after 6 months of tadalafil therapy could be tested in a prospective and randomized trial to assess whether tadalafil provides clinical benefit in patients with heart failure and preserved ejection fraction. Previous studies have investigated the effects of PDE5i on cardiac function, reporting controversial results. In 44 patients with stable systolic heart failure randomly assigned to placebo or sildenafil, Guazzi et al. demonstrated that PDE5 inhibition with sildenafil significantly improves left ventricular diastolic function properties [22]. In contrast, in the subsequent Relax trial by Redfield et al., including 216 patients with heart failure with preserved ejection fraction, sildenafil showed no beneficial effects on exercise capacity and left ventricular diastolic function [23]. Similarly, Andersen et al. reported that sildenafil treatment for 9 weeks did not show any significant improvements in the predetermined primary endpoints (LV filling pressure at rest or during exercise) in 70 patients with diastolic dysfunction and preserved ejection fraction after acute myocardial infarction. In addition, Andersen and coworkers did not detect any significant improvement in diastolic function, as assessed by Doppler echocardiography, following the treatment with sildenafil [24]. Differently, in a 12-week, randomized, double-blind, placebo-controlled trial, Kim et al. reported that PDE 5 inhibition with udenafil ameliorates systolic and diastolic function of the left ventricle along with exercise capacity in patients with chronic heart failure with reduced ejection fraction [25]. A subsequent meta-analysis performed by De Vecchis et al. found that PDE5 inhibitors improved clinical outcomes, exercise capacity, and pulmonary hemodynamics in the RCTs of patients (n = 555) with heart failure and reduced left ventricular ejection fraction [26]. In contrast, in the RCTs of patients with heart failure with preserved left ventricular ejection fraction (n = 373), no benefit was shown from PDE5i use regarding all of the investigated endpoints. Again, in a randomized placebo-controlled study, Liu et al. found that treatment with sildenafil for 12 weeks in 52 patients with heart failure with preserved ejection fraction and predominantly isolated post-capillary pulmonary hypertension did not affect the cardiac structure, cardiac function, exercise response, or quality of life [27]. More recently, Belyavskiy et al. investigated 50 patients with heart failure with preserved ejection fraction and combined pre- and post-capillary pulmonary hypertension [28]. In the 30 patients treated with sildenafil for 6 months, the authors observed improvements in exercise capacity, pulmonary hemodynamic parameters, and right ventricular function. No changes occurred in the control group [29].

One explanation for the seemingly contradictory findings in the above studies may be due to the differences in the assessment methods of the effects of the drugs. It is also possible that a lack of homogeneity in the characteristics of the study populations in terms of etiology, comorbidities, and therapies associated with PDE5 inhibitors may have contributed. The efficacy of PDE5 inhibitors might differ according to the pathological process under study. A single class of drugs cannot prove effective in all clinical scenarios. PDE5 inhibitors demonstrated clinical benefit in patients with heart failure with reduced ejection fraction. In the preserved ejection fraction setting, PDE5 inhibitors may only benefit select patients with reversible pulmonary hypertension and right ventricular systolic dysfunction. Additionally, further differences in cardiovascular response to treatment could be found in the different genetic backgrounds. Indeed, Manca P. et al. demonstrated that the impact of heart failure treatment on the right heart chamber could be influenced by different genetic backgrounds, determining heterogenic results [30]. Given the small numbers of patients enrolled in the various studies, we cannot exclude a play of chance. The PASSION trial evaluating the impact of tadalafil on clinical endpoints in patients with combined pre- and post-capillary pulmonary hypertension due to heart failure with preserved ejection fraction is ongoing, under the direction of Dr. M. Hoeper, Hanover, and Dr. S. Rosenkranz, Cologne, and will provide relevant information with potential implications for the clinical use of PDE5 inhibition in patients with heart failure and preserved ejection fraction [31].

Several study limitations should be acknowledged. Because of the small sample size, this study should be interpreted as a preliminary investigation; the precision for estimation of the magnitude of effects of tadalafil is far too low. Given the observational nature of the study, no sample size estimation was calculated. The probands were in control of themselves after 6 months of treatment; thus, a randomized control design of the study was not conceivable. Therefore, the possible evidence of causality between tadalafil therapy and improvement in diastolic function should be considered. The enrolled population was limited to prostate cancer patients undergoing robot-assisted radical prostatectomy at our tertiary urologic center; thus, no information about the impact of more invasive, conventional surgical treatment was explored. Possible additional beneficial effects on the cardiovascular system could be assumed secondary to a less invasive treatment.

5. Conclusions

In conclusion, the efficacy of tadalafil in improving some echocardiographic parameters of left ventricular diastolic function observed in our study must be interpreted as hypothesis-generating. We recommend that the role of tadalafil in the prevention and treatment of diastolic dysfunction should be considered for future studies.

Author Contributions: Conceptualization, F.C. and R.E.; methodology, L.F.; formal analysis, V.C.; investigation, C.S.; data curation, B.B.; writing—original draft preparation, D.A.; writing—review and editing, C.I.; supervision, T.F. All authors have read and agreed to the published version of the manuscript.

Funding: This research received no external funding.

Institutional Review Board Statement: Research Ethics Board of "Federico II" University of Naples, n.316/20.

Informed Consent Statement: Informed consent was obtained from all subjects involved in the study.

Acknowledgments: In memory of the late Maurizio Galderisi who helped in the conceptualization of this project.

Conflicts of Interest: The authors declare no conflict of interest.

References

1. Ghofrani, H.A.; Osterloh, I.H.; Grimminger, F. Sildenafil: From angina to erectile dysfunction to pulmonary hypertension and beyond. *Nat. Rev. Drug Discov.* **2006**, *5*, 689–702. [CrossRef] [PubMed]
2. Huang, S.A.; Lie, J.D. Phosphodiesterase-5 (PDE5) Inhibitors in the Management of Erectile Dysfunction. *J. Formul. Manag.* **2013**, *38*, 407–419.
3. Burnett, A.L. The Role of Nitric Oxide in Erectile Dysfunction: Implications for Medical Therapy. *J. Clin. Hypertens.* **2006**, *8*, 53–62. [CrossRef] [PubMed]
4. Salloum, F.N.; Takenoshita, Y.; Ockaili, R.A.; Daoud, V.P.; Chou, E.; Yoshida, K.; Kukreja, R.C. Sildenafil and vardenafil but not nitroglycerin limit myocardial infarction through opening of mitochondrial K(ATP) channels when administered at reperfusion following ischemia in rabbits. *J. Mol. Cell Cardiol.* **2007**, *42*, 453–458. [CrossRef] [PubMed]
5. Lee, K.H.; Kwon, S.J.; Woo, J.S.; Lee, G.J.; Lee, S.R.; Jang, H.H.; Kim, H.S.; Kim, J.W.; Park, H.K.; Cho, K.S.; et al. Effects of sildenafil on nanostructural and nanomechanical changes in mitochondria in an ischaemia-reperfusion rat model. *Clin. Exp. Pharmacol. Physiol.* **2014**, *41*, 763–768. [CrossRef]
6. Mátyás, C.; Németh, B.T.; Oláh, A.; Török, M.; Ruppert, M.; Kellermayer, D.; Barta, B.A.; Szabó, G.; Kökény, G.; Horváth, E.M.; et al. Prevention of the development of heart failure with preserved ejection fraction by the phosphodiesterase-5A inhibitor vardenafil in rats with type 2 diabetes. *Eur. J. Heart Fail.* **2017**, *19*, 326–336. [CrossRef]

7. Frankenreiter, S.; Bednarczyk, P.; Kniess, A.; Bork, N.I.; Straubinger, J.; Koprowski, P.; Wrzosek, A.; Mohr, E.; Logan, A.; Murphy, M.P.; et al. cGMP-Elevating Compounds and Ischemic Conditioning Provide Cardioprotection Against Ischemia and Reperfusion Injury via Cardiomyocyte-Specific BK Channels. *Circulation* **2017**, *136*, 2337–2355. [CrossRef]
8. Imai, Y.; Kariya, T.; Iwakiri, M.; Yamada, Y.; Takimoto, E. Sildenafil ameliorates right ventricular early molecular derangement during left ventricular pressure overload. *PLoS ONE* **2018**, *13*, e0195528. [CrossRef]
9. Noale, M.; The Pros-IT CNR study group; Maggi, S.; Artibani, W.; Bassi, P.F.; Bertoni, F.; Bracarda, S.; Conti, G.N.; Corvò, R.; Gacci, M.; et al. Pros-IT CNR: An Italian prostate cancer monitoring project. *Aging Clin. Exp. Res.* **2017**, *29*, 165–172. [CrossRef]
10. Kong, Q.; Blanton, R.M. Protein kinase G I and heart failure: Shifting focus from vascular unloading to direct myocardial antiremodeling effects. *Circ. Heart Fail.* **2013**, *6*, 1268–1283. [CrossRef]
11. Aydogdu, O.; Gokce, M.I.; Burgu, B.; Baltaci, S.; Yaman, O. Tadalafil rehabilitation therapy preserves penile size after bilateral nerve sparing radical retropubic prostatectomy. *Int. Braz J. Urol.* **2011**, *37*, 336–346. [CrossRef] [PubMed]
12. Lang, R.M.; Badano, L.P.; Mor-Avi, V.; Afilalo, J.; Armstrong, A.; Ernande, L.; Lachskampf, F.A.; Foster, E.; Goldstein, S.A.; Kuznestova, T.; et al. Recommendations for cardiac chamber quantification by echocardiography in adults: An update from the American Society of Echocardiography and the European Association of Cardiovascular Imaging. *J. Am. Soc. Echocardiogr.* **2015**, *28*, 1–39. [CrossRef] [PubMed]
13. Nagueh, S.F.; Smiseth, O.A.; Appleton, C.P.; Byrd, B.F., 3rd; Dokainish, H.; Edvardsen, T.; Flachskampf, F.A.; Gillebert, T.C.; Klein, A.L.; Lancellotti, P.; et al. Recommendations for the Evaluation of Left Ventricular Diastolic Function by Echocardiography: An Update from the American Society of Echocardiography and the European Association of Cardiovascular Imaging. *J. Am. Soc. Echocardiogr.* **2016**, *29*, 277–314. [CrossRef] [PubMed]
14. Santoro, C.; Esposito, R.; Lembo, M.; Sorrentino, R.; De Santo, I.; Luciano, F.; Casciano, O.; Giuliano, M.; De Placido, S.; Trimarco, B.; et al. Strain-oriented strategy for guiding cardioprotection initiation of breast cancer patients experiencing cardiac dysfunction. *Eur. Heart J. Cardiovasc. Imaging* **2019**, *20*, 1345–1352. [CrossRef]
15. Alcidi, G.M.; Esposito, R.; Evola, V.; Santoro, C.; Lembo, M.; Sorrentino, R.; Lo Iudice, F.; Borgia, F.; Novo, G.; Trimarco, B.; et al. Normal reference values of multilayer longitudinal strain according to age decades in a healthy population: A single-centre experience. *Eur. Heart J. Cardiovasc. Imaging* **2017**, *19*, 1390–1396. [CrossRef]
16. Sorrentino, R.; Esposito, R.; Santoro, C.; Vaccaro, A.; Cocozza, S.; Scalamogna, M.; Lembo, M.; Luciano, F.; Santoro, A.; Trimarco, B.; et al. Practical Impact of New Diastolic Recommendations on Noninvasive Estimation of Left Ventricular Diastolic Function and Filling Pressures. *J. Am. Soc. Echocardiogr.* **2020**, *33*, 171–181. [CrossRef]
17. Rudski, L.G.; Lai, W.W.; Afilalo, J.; Hua, L.; Handschumacher, M.D.; Chandrasekaran, K.; Solomon, S.D.; Louie, E.K.; Schiller, N.B. Guidelines for the echocardiographic assessment of the right heart in adults: A report from the American Society of Echocardiography endorsed by the European Association of Echocardiography, a registered branch of the European Society of Cardiology, and the Canadian Society of Echocardiography. *J. Am. Soc. Echocardiogr.* **2010**, *23*, 685–713.
18. Armstrong, P.W.; Pieske, B.; Anstrom, K.J.; Ezekowitz, J.; Hernandez, A.F.; Butler, J.; Lam, C.S.P.; Ponikowski, P.; Voors, A.A.; Jia, G.; et al. Vericiguat in Patients with Heart Failure and Reduced Ejection Fraction. *N. Engl. J. Med.* **2020**, *382*, 1883–1893. [CrossRef]
19. McDonagh, T.A.; Metra, M.; Adamo, M.; Gardner, R.S.; Baumbach, A.; Böhm, M.; Burri, H.; Butler, J.; Čelutkienė, J.; Chioncel, O.; et al. 2021 ESC Guidelines for the diagnosis and treatment of acute and chronic heart failure: Developed by the Task Force for the Diagnosis and Treatment of Acute and Chronic Heart Failure of the European Society of Cardiology (ESC) With the Special Contribution of the Heart Failure Association (HFA) of the ESC. *Eur. Heart J.* **2021**, *42*, 3599–3726. [CrossRef]
20. Kane, G.C.; Karon, B.L.; Mahoney, D.W.; Redfield, M.M.; Roger, V.L.; Burnett, J.C.; Jacobsen, S.; Rodeheffer, R.J. Progression of Left Ventricular Diastolic Dysfunction and Risk of Heart Failure. *J. Am. Med. Assoc.* **2011**, *306*, 856–863. [CrossRef]
21. Echouffo-Tcheugui, J.B.; Erqou, S.; Butler, J.; Yancy, C.W.; Fonarow, G.C. Assessing the Risk of Progression from Asymptomatic Left Ventricular Dysfunction to Overt Heart Failure: A Systematic Overview and Meta-Analysis. *JACC Heart Fail.* **2016**, *4*, 237–248. [CrossRef] [PubMed]
22. Guazzi, M.; Vicenzi, M.; Arena, R.; Guazzi, M.D. PDE5 inhibition with sildenafil improves left ventricular diastolic function, cardiac geometry, and clinical status in patients with stable systolic heart failure: Results of a 1-year, prospective, randomized, placebo-controlled study. *Circ. Heart Fail.* **2011**, *4*, 8–17. [CrossRef] [PubMed]
23. Redfield, M.M.; Chen, H.H.; Borlaug, B.A.; Semigran, M.J.; Lee, K.L.; Lewis, G.; LeWinter, M.M.; Rouleau, J.L.; Bull, D.A.; Mann, D.L.; et al. RELAX Trial. Effect of phosphodiesterase-5 inhibition on exercise capacity and clinical status in heart failure with preserved ejection fraction: A randomized clinical trial. *J. Am. Med. Assoc.* **2013**, *309*, 1268–1277. [CrossRef]
24. Andersen, M.J.; Ersbøll, M.; Axelsson, A.; Gustafsson, F.; Hassager, C.; Køber, L.; Borlaug, B.A.; Boesgaard, S.; Skovgaard, L.T.; Møller, J.E. Sildenafil and diastolic dysfunction after acute myocardial infarction in patients with preserved ejection fraction: The Sildenafil and Diastolic Dysfunction After Acute Myocardial Infarction (SIDAMI) trial. *Circulation* **2013**, *127*, 1200–1208. [CrossRef] [PubMed]
25. Kim, K.-H.; Kim, H.-K.; Hwang, I.-C.; Cho, H.-J.; Je, N.; Kwon, O.-M.; Choi, S.-J.; Lee, S.-P.; Kim, Y.-J.; Sohn, D.-W. PDE 5 inhibition with udenafil improves left ventricular systolic/diastolic functions and exercise capacity in patients with chronic heart failure with reduced ejection fraction; A 12-week, randomized, double-blind, placebo-controlled trial. *Am. Heart J.* **2015**, *169*, 813–822.e3. [CrossRef] [PubMed]

26. De Vecchis, R.; Cesaro, A.; Ariano, C.; Giasi, A.; Cioppa, C. Phosphodiesterase-5 Inhibitors Improve Clinical Outcomes, Exercise Capacity and Pulmonary Hemodynamics in Patients with Heart Failure with Reduced Left Ventricular Ejection Fraction: A Meta-Analysis. *J. Clin. Med. Res.* **2017**, *9*, 488–498. [CrossRef]
27. Liu, L.C.; Hummel, Y.M.; van der Meer, P.; Berger, R.M.; Damman, K.; van Veldhuisen, D.J.; Voors, A.A.; Hoendermis, E.S. Effects of sildenafil on cardiac structure and function, cardiopulmonary exercise testing and health-related quality of life measures in heart failure patients with preserved ejection fraction and pulmonary hypertension. *Eur. J. Heart Fail.* **2017**, *19*, 116–125. [CrossRef]
28. Belyavskiy, E.; Ovchinnikov, A.; Potekhina, A.; Ageev, F.; Edelmann, F. Phosphodiesterase 5 inhibitor sildenafil in patients with heart failure with preserved ejection fraction and combined pre- and postcapillary pulmonary hypertension: A randomized open-label pilot study. *BMC Cardiovasc. Disord.* **2020**, *20*, 1–15. [CrossRef]
29. *Phosphodiesterase-5 Inhibition in Patients with Heart Failure with Preserved Ejection Fraction and Combined Post- and Pre-Capillary Pulmonary Hypertension (PASSION), A Randomized, Placebo-Controlled, Double-Blind, Parallel Group, Multi-Center, Phase III Trial; EudraCT number 2017-003688-37*; European Medicines Agency: Amsterdam, The Netherlands, 2018.
30. Manca, P.; Cannatà, A.; Nuzzi, V.; Bromage, D.I.; Varrà, G.G.; Rossi, M.; Ferro, M.D.; Paldino, A.; Gigli, M.; Barbati, G.; et al. Prevalence and Evolution of Right Ventricular Dysfunction Among Different Genetic Backgrounds in Dilated Cardiomyopathy. *Can. J. Cardiol.* **2021**, *37*, 1743–1750. [CrossRef]
31. Cameli, M.; Miglioranza, M.H.; Magne, J.; Mandoli, G.E.; Benfari, G.; Ancona, R.; Sibilio, G.; Luksic, V.R.; Dejan, D.; Griseli, L.; et al. Multicentric Atrial Strain COmparison between Two Different Modalities: MASCOT HIT Study. *Diagnostics* **2020**, *10*, 946. [CrossRef]

applied sciences

MDPI

Article

A Novel Speckle-Tracking Echocardiography Derived Parameter That Predicts Clinical Worsening in Children with Pulmonary Arterial Hypertension

Iolanda Muntean [1,*], Mihaela Melinte [2], Amalia Făgărășan [1], Carmen Corina Șuteu [3] and Rodica Togănel [1]

1 Clinic of Paediatric Cardiology, Emergency Institute for Cardiovascular and Transplant Diseases of Târgu Mureș, University of Medicine Pharmacy Sciences and Technology "George Emil Palade", 540142 Târgu Mureș, Romania; amalia.fagarasan@yahoo.com (A.F.); rodicatoganel@yahoo.com (R.T.)
2 Laboratory Department, Emergency Institute for Cardiovascular and Transplant Diseases of Târgu Mureș, 540136 Târgu Mureș, Romania; mihaela_im@yahoo.com
3 Clinic of Paediatric Cardiology, Emergency Institute for Cardiovascular and Transplant Diseases of Târgu Mureș, 540136 Târgu Mureș, Romania; suteucarmen@yahoo.com
* Correspondence: iolanda.muntean@gmail.com; Tel.: +40-744-372276

Abstract: Pulmonary arterial hypertension is a severe, progressive disease in children, that causes right ventricular dysfunction over time. Tissue motion annular displacement is a novel speckle-tracking derived echocardiographic parameter used in assessing ventricular function. The aim of our study was to determine the prognostic value of this echocardiographic parameter in children with pulmonary arterial hypertension. We conducted a case-control study by assessing twenty children with pulmonary arterial hypertension (idiopathic or secondary) and twenty age- and sex-matched controls, using clinical (WHO functional class, 6-min walking test), laboratory (brain natriuretic peptide level) and echocardiographic parameters (conventional and speckle-tracking derived tissue motion annular displacement) at enrolment and after one year of follow-up. According to their WHO functional class altering after one year, the pulmonary arterial hypertension patients were divided into two groups: non-worsening (eleven) and worsening (nine). The conventional echocardiographic parameters and all measured tricuspid tissue motion annular displacement indices (lateral, septal, midpoint and midpoint fractional displacement—TMADm%) were significantly lower in both pulmonary arterial hypertension groups (non-worsening and worsening) compared to controls. Comparing the worsening and non-worsening groups, only the TMADm% and brain natriuretic peptide level was significantly lower in worsening in comparison with non-worsening pulmonary arterial hypertension children ($p = 0.010$ and $p = 0.018$, respectively). In receiver-operating characteristic curve analysis, we found a cut-off value of 16.15% for TMADm% and a cut-off value of 34.35 pg/mL for the brain natriuretic peptide level that can predict worsening in pulmonary arterial hypertension children. In conclusion, tricuspid annulus midpoint fractional displacement, an angle-dependent speckle-tracking derived parameter, could be a good additional parameter in the assessment of the longitudinal right ventricular systolic function and in prediction of clinical worsening in children with pulmonary arterial hypertension.

Keywords: right ventricle; speckle-tracking; tricuspid tissue motion annular displacement; pulmonary arterial hypertension; children

Citation: Muntean, I.; Melinte, M.; Făgărășan, A.; Șuteu, C.C.; Togănel, R. A Novel Speckle-Tracking Echocardiography Derived Parameter That Predicts Clinical Worsening in Children with Pulmonary Arterial Hypertension. *Appl. Sci.* **2022**, *12*, 5494. https://doi.org/10.3390/app12115494

Academic Editor: Julio Garcia Flores

Received: 13 March 2022
Accepted: 26 May 2022
Published: 28 May 2022

1. Introduction

Pulmonary arterial hypertension (PAH), included in the first group of pulmonary hypertensions (Nice, 2018) [1], is a severe, progressive disease in children, that causes pulmonary vascular remodeling and right ventricular (RV) dysfunction over time.

Transthoracic echocardiography is the most accessible, non-invasive diagnostic tool for the initial assessment and clinical follow-up of children with PAH [2–4]. In the current guidelines, the importance of the multiparametric approach in the echocardiographic assessment of these children is emphasized [3].

According to the risk stratification model implemented by the 6th World Symposium on Pulmonary Hypertension—Pediatric Task Force in 2018 (Nice), and by the updated consensus document of the "European Pediatric Pulmonary Vascular Disease Network", some echocardiographic parameters have been suggested as risk factors and predictors of outcome such as right atrial (RA) and right ventricular (RV) enlargement, reduced left ventricular (LV) size, increased RV/LV ratio, reduced tricuspid annular plane systolic excursion (TAPSE), increased systolic/diastolic (S/D) ratio, pulmonary artery acceleration time, low RV fractional area changes (FAC) and pericardial effusion [3,4]. However, because these parameters have only level of evidence C (due to lack of pediatric data) and because of the significant operator and interpretation variability of some echocardiographic parameters, the need for identification of more valid, easy-to-determine predictors and treatment goals in pediatric patients with PAH is emphasized [3].

Tissue motion annular displacement (TMAD) is a speckle-tracking derived echocardiographic parameter in which the annular tissue is tracked toward the ventricular apex. This angle-independent echocardiographic parameter has been used in assessing left (TMAD of the mitral valve) [5–8] and right (TMAD of the tricuspid valve) [9–11] ventricular function. However, its usefulness in the assessment of the right ventricular function in PAH children has not been studied so far.

The aim of our study was to determine the prognostic value of TMAD of the tricuspid valve in children with PAH.

2. Materials and Methods

2.1. Study Population

We have performed a clinical observational study on forty patients: twenty children with PAH (five idiopathic PAH, twelve PAH secondary to ventricular septal defect and three PAH secondary to truncus arteriosus communis) and twenty age- and sex-matched healthy children as a control group. After enrolment, all forty patients have undergone complete physical examination including 6 min walking test, standard and advanced echocardiographic examination (speckle tracking derived TMAD) and blood sampling for brain natriuretic peptide and were repeated after one year of follow-up. According to their WHO functional class altering after one year, the patients were divided into two groups: *worsening* and *non-worsening*. The patients were recruited at the same time from baseline and all the patients have the same 1-year follow-up. There was not any follow-up lost. We followed the methods of Muntean et al. [12]. The general characteristics of the patients involved in the present study are presented in Table 1.

The PAH was stated according to the NICE 2018 definition [4], and all PAH children were on pulmonary vasodilator medication (mono- or combined drug). In children from the control group, who were investigated for fatigability or chest pain, any cardiac disease was ruled out.

Exclusion criteria included patients with other causes of PAH. Patients older than 18 years or those with arrhythmia were excluded from the present study.

Table 1. General characteristics of the children from the study groups (control, non-worsening PAH children and worsening PAH children).

Variable	Control (n = 20)	Non-Worsening PAH Children (n = 11)	Worsening PAH Children (n = 9)	p_1 (nw vs. c)	p_2 (w vs. c)	p_3 (w vs. nw)
Age (y)	10.72 ± 3.65	10.59 ± 3.33	10.15 ± 3.80	1.000	1.000	1.000
Male (%)	9 (45%)	3 (27.27%)	4 (44.44%)	0.451	1.000	0.642
H (m)	1.45 ± 0.20	1.32 ± 0.15	1.31 ± 0.23	0.239	0.267	1.000
W (kg)	40.60 ± 18.67	26.95 ± 8.45	29.21 ± 10.82	0.060	0.197	1.000
BSA (m^2) BMI (m/kg^2)	1.26 ± 0.37 17.95 ± 3.53	0.98 ± 0.20 15.00 ± 2.33	1.02 ± 0.28 16.16 ± 2.04	0.077 0.035	0.228 0.422	1.000 1.000
6MWD (m)	473.05 ± 46.51	419.18 ± 54.50	406 ± 69.50	0.036	0.012	1.000
WHO Functional class 1/2/3/4—at baseline	NA	0/8/3/0 2.00 (2.00, 3.00)	1/4/3/1 2.44 ± 0.88	NA	NA	0.656
WHO Functional class 1/2/3/4—after 1 year	NA	0/10/1/0 2.00 (2.00, 2.00)	0/1/4/4 3.33 ± 0.70	NA	NA	0.001
BNP level (pg/mL)	9.00 (9.00, 15.60)	17.90 (10.60, 56.40)	53.38 ± 27.14	0.243	0.0001	0.018

Data are presented as means ± SD or median (25th, 75th percentiles): BMI, body mass index; BNP, brain natriuretic peptide; BSA, body surface area; c, control; H, height; HR, heart rate; nw, non-worsening; PAH, pulmonary arterial hypertension; p_1, non-worsening group vs. control; p_2, worsening group vs. control; p_3, worsening group vs. non-worsening; W, weight; w, worsening; y, year; WHO, World Health Organization; 6MWD, 6-min walk distance; variables are expressed as means ± SD, medians (25th, 75th percentiles) or as numbers (percentages).

2.2. *Standard Echocardiographic Examination*

The echocardiographic examination was performed by an iE33 (Philips Medical Systems, Best, The Netherlands) ultrasound system using an S5-1 transducer, by a single sonographer and minimum three beats were stored. Depth, width and frame rate (60–100/s) were adjusted for an accurate speckle-tracking postprocessing.

During the echocardiographic examination we have measured the following parameters TAPSE, RV-FAC, RV-S/D ratio, RV-MPI (myocardial performance index), RA area, tissue Doppler derived systolic myocardial velocity (S'), left ventricular eccentricity index (LV-EI) as recommended by the current guidelines [2,13,14].

2.3. *TMAD Data Acquisition and Analysis*

TMAD of the tricuspid valve was measured in the RV focused apical four-chamber view, using the 2-dimensional speckle tracking technique. The speckle tracking analysis was performed offline using the QLAB software (QLAB 15.0, Cardiac motion quantification software, Philips Medical Systems, Best, The Netherlands). Three points were selected in a diastolic frame: insertion of the anterior leaflet and the septal leaflet, respectively, into the tricuspid annulus and the right ventricular apex. Further, the lateral (TMADlat) and septal (TMADsept) tricuspid annular points longitudinal displacement were tracked during the cardiac cycle. Midpoint displacement (TMADmid) of the tricuspid valve annulus was also computed by the software. Further, tricuspid annulus midpoint fractional displacement (TMADm%) also known as right ventricular longitudinal shortening fraction (RV-LSF) was calculated automatically by the software, as the maximum displacement of the midpoint throughout the cardiac cycle, as follows: (end diastolic RV length − end systolic RV length) × 100/(end diastolic RV length) (Figure 1) [15].

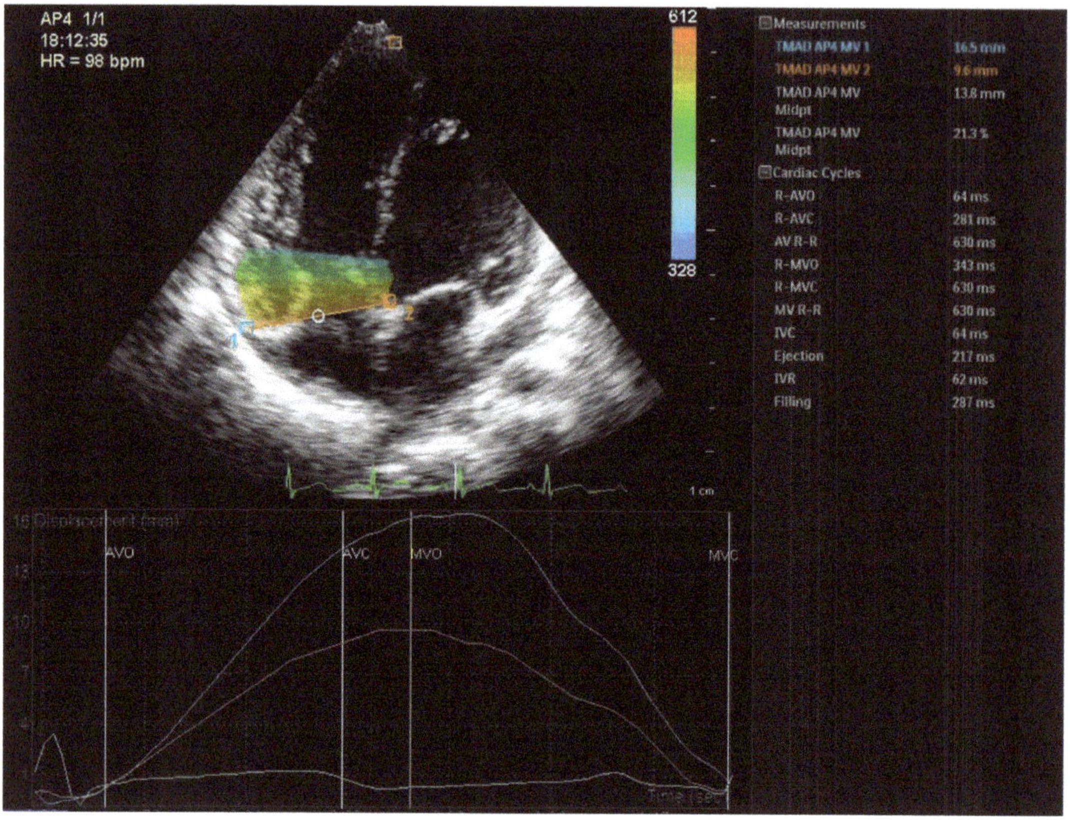

Figure 1. Speckle-tracking derived tissue annular motion displacement of the tricuspide valve assessment. Three points were selected: insertion of the anterior tricuspid leaflet (1), the septal tricuspid leaflet (2), and the right ventricular apex.

2.4. Statistical Analysis

Statistical analysis was performed using SPSS version 20 (IBM SPSS STATISTICS 20). Data were categorized as nominal or quantitative variables. Nominal variables were expressed as numbers or percentages. Quantitative variables were expressed by mean $\pm$ standard deviation or median and percentiles (25; 75%), whenever appropriate. A Kolmogorov–Smirnov test was used in order to test for normality of distribution of quantitative variables. Differences between the mean or median between two groups were analyzed using the *t*-test, or Mann–Whitney test when appropriate. Differences between three groups with normally distributed variables were analyzed using general linear model test with Bonferroni post-hoc processing. Receiver-operating characteristic (ROC) curves were constructed, and areas under curve were calculated. Sensitivities and specificities were determined for the ability to identify worsening PAH children. A *p*-value of <0.05 was considered statistically significant.

3. Results

3.1. General Information

A total of forty children were included in the present study, twenty with PAH (eleven in the non-worsening group and nine in the worsening group) and twenty control children. As summarized in Table 1, there were no statistically significant differences between the three groups regarding age, sex, height, weight and body surface area.

Regarding 6MWD, we found a progressive decrease in this parameter in the three groups, with the largest value in the control group and the lowest in the worsening PAH group. Further, we observed that the BNP level was higher in both PAH groups in comparison to the control group, with the largest value in the worsening group. We found a statistically significant difference between worsening group and the two other groups.

3.2. Conventional Echocardiographic Parameters

Table 2 shows comparison of echocardiographic parameters between the three study groups (control, non-worsening PAH children, worsening PAH children).

Table 2. Comparison of echocardiographic parameters between the study groups (control, non-worsening PAH children and worsening PAH children).

Variable	Control (n = 18)	Non-Worsening PAH Children (n = 10)	Worsening PAH Children (n = 7)	p_1 (nw vs. c)	p_2 (w vs. c)	p_3 (w vs. nw)
TR (mmHg)	NA	89.54 ± 32.05	95.44 ± 12.73	NA	NA	0.585
PAPm (mmHg)	NA	61.900 ± 20.19	57.00 (51.00, 59.00)	NA	NA	0.766
Indexed RV-ED area (cm^2)	13.31 ± 5.03	14.62 ± 3.88	17.55 ± 6.30	1.000	0.133	0.623
Indexed RV-ES area (cm^2)	6.61 (5.54, 8.58)	9.98 ± 3.99	11.78 ± 4.68	0.214	0.017	0.882
RV-FAC	0.45 ± 0.05	0.33 ± 0.13	0.33 ±0.06	0.002	0.007	1.000
RV MPI	0.12 ± 0.07	0.37 ± 0.18	0.33 (0.28, 0.70)	0.0001	0.0001	0.492
TAPSE (cm)	2.16 ± 0.46	1.62 ± 0.47	1.62 ± 0.53	0.015	0.023	1.000
RV S (ms)	355.63 ± 33.76	377.27 ± 33.23	397.22 ± 55.72	0.457	0.037	0.803
RV D (ms)	397.12 ± 97.63	283.35 ± 69.74	260.00 (250.58, 347.20)	0.007	0.056	1.000
RV S/D ratio	0.85 (0.77, 1.12)	1.39 ± 0.31	1.38 ± 0.33	0.0001	0.001	1.000
Indexed RA area (cm^2)	8.38 ± 1.35	10.67 (9.60, 13.79)	11.38 (10.20, 16.27)	0.945	0.018	0.268
Diam 1 LV (cm)	3.94 ± 0.65	4.24 ± 0.63	4.05 ± 0.46	0.602	1.000	1.000
Diam 2 LV (cm)	3.91 ± 0.55	3.12 ± 0.40	2.90 ± 0.59	0.001	0.0001	1.000
LV-EI	1.00 ± 0.07	1.26 (1.19, 1.63)	1.44 ± 0.29	0.0001	0.0001	1.000
S′ (cm/s)	13.43 ± 1.60	10.92 ± 2.07	10.44 ± 2.68	0.006	0.002	1.000
TMAD lat (mm)	19.08 ± 3.88	14.90 ± 2.37	11.64 ± 2.84	0.006	0.0001	0.106
TMAD sept (mm)	12.29 ± 2.25	8.51 ± 1.82	7.63 ± 2.09	0.0001	0.0001	1.000
TMADmid (mm)	16.38 ± 2.96	12.15 ± 2.13	10.44 ± 3.59	0.001	0.0001	0.603
TMADm%	23.70 ± 3.62	18.13 ± 2.87	13.51 ± 2.84	0.0001	0.0001	0.010

Data are presented as means ± SD or median (25th, 75th percentiles): c, control; D, diastole, ED, end-diastolic; EI, eccentricity index; ES, end-systolic; FAC, fractional area changes; IVS, interventricular septum; LV, left ventricle; MPI, myocardial performance index; NA, no available; nw, non-worsening; PAH, pulmonary arterial hypertension; PAPm, medium pulmonary arterial pressure; p_1, non-worsening group vs. control; p_2, worsening group vs. control; p_3, worsening group vs. non-worsening; RA, right atrium; RV, right ventricle; S, systole; S′, tricuspid annular systolic velocity; TAPSE, tricuspid annular plane systolic excursion; TMAD, tissue motion annular displacement; TR, tricuspid regurgitation; w, worsening.

Regarding conventional echocardiographic parameters, significant differences were observed in RV-FAC, RV-MPI, RV S/D, S′, LV-EI between non-worsening PAH children and control group. We performed the receiver-operator characteristic analysis for all these parameters, and the reference group was the non-worsening PAH group. The ROC curves are presented in Figure 2.

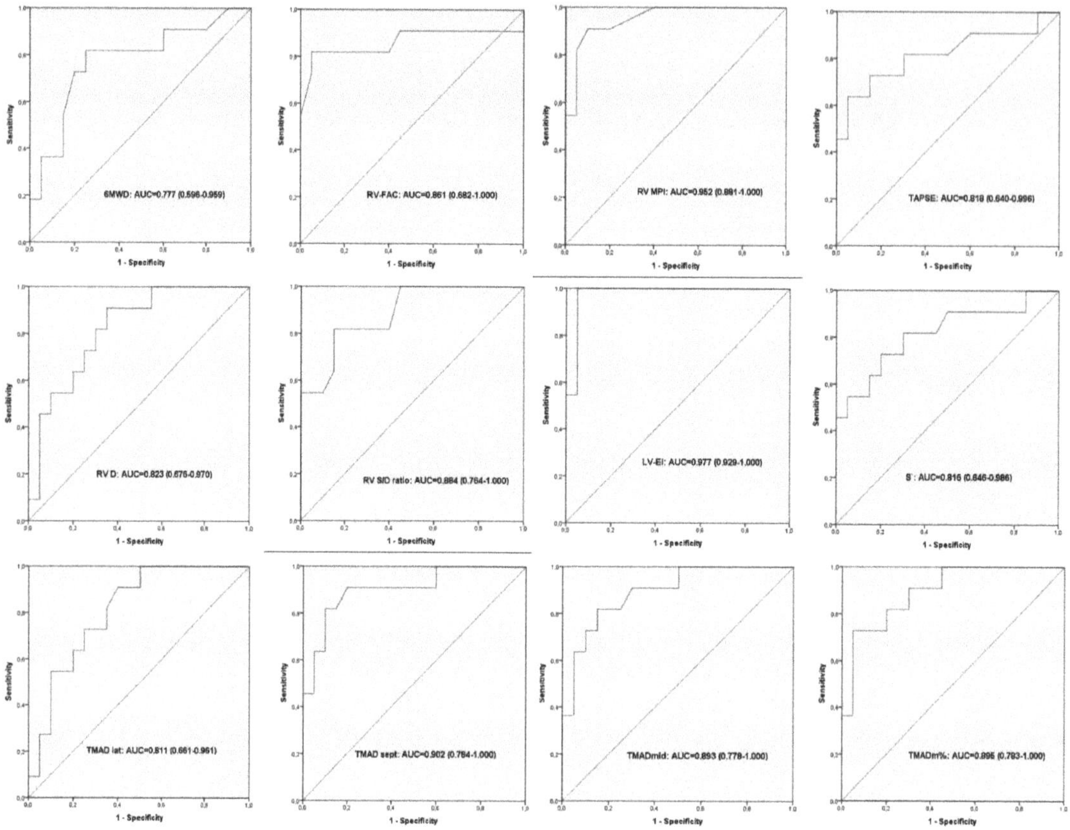

Figure 2. Receiver-operator characteristic analysis. ROC curves result from all the variables (echocardiography and 6MWD) that showed significant difference between controls and non-worsening PAH group.

The same significant differences were observed in RV-FAC, RV-MPI, RV S/D, S′, LV-EI between the worsening group and control group. In addition, comparing these two last groups, we also found a statistical difference in the indexed RV and RA area, explained by the fact that in the worsening group, the right ventricle was more enlarged than in the non-worsening group (Table 2). We have also performed the receiver-operator characteristic analysis for all these parameters, and the reference group was the worsening PAH group. The ROC curves are presented in Figure 3.

However, comparing the worsening and non-worsening PAH groups, we found no statistically significant differences regarding conventional echocardiographic parameters (Table 2).

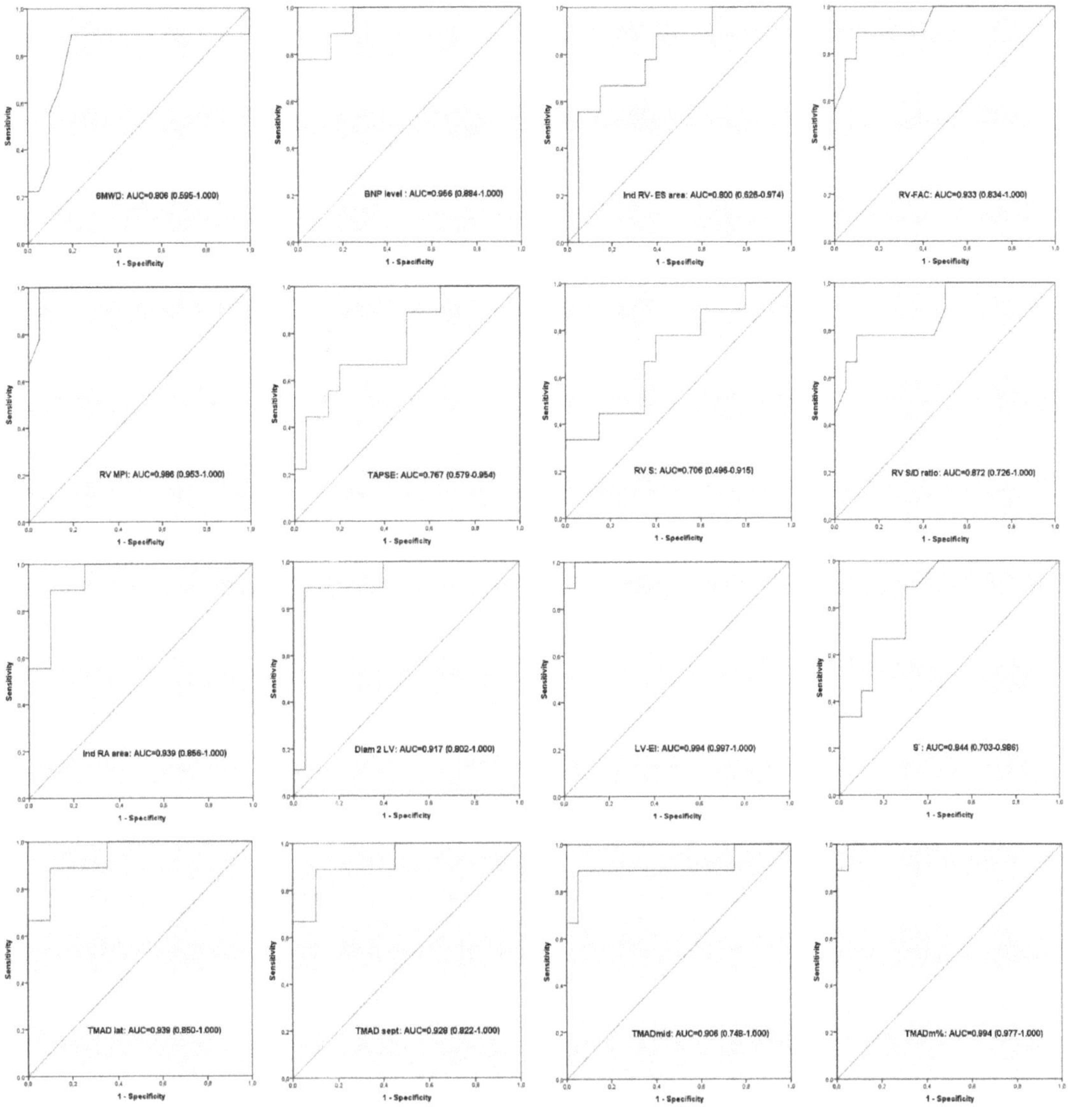

Figure 3. Receiver-operator characteristic analysis. ROC curves result from all the variables (echocardiography, BNP level and 6MWD) that showed significant difference between controls and worsening PAH group.

3.3. TMAD Analysis

Considering the speckle-tracking derived TMAD of tricuspid valve, all measured TMAD indices (TMADlat, TMADsept, TMADmid and TMADm%) were significantly decreased in the non-worsening group as well as in the worsening group in comparison with the control group. Additionally, in all three groups, we found a descendent distribution of the TMAD values from the lateral to mid and to the septal point (Figure 4).

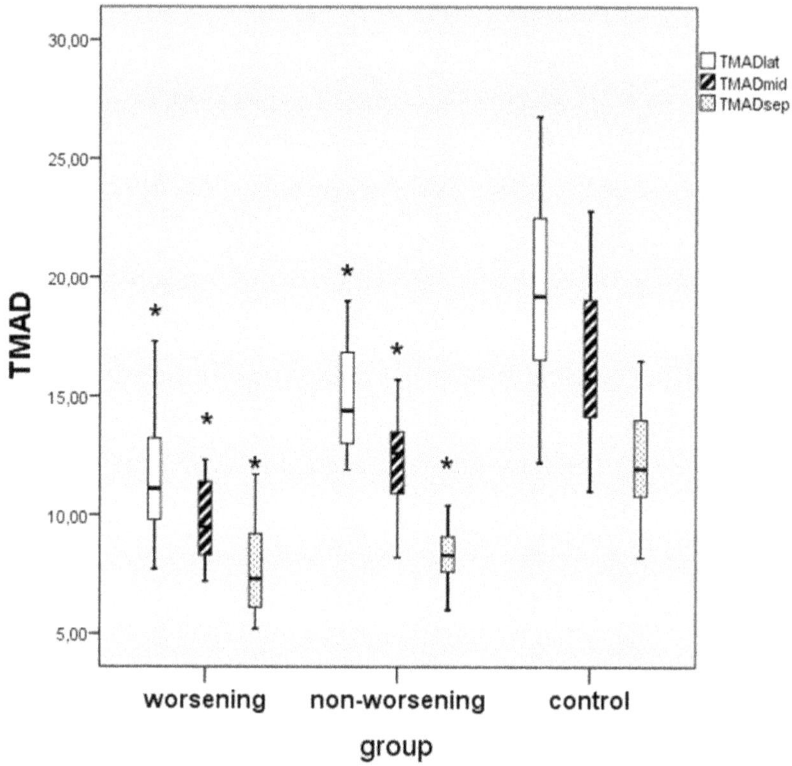

Figure 4. Comparison between worsening, non-worsening PAH children and controls regarding TMAD of tricuspid valve values in three points: lateral, midpoint and septal. * worsening or non-worsening PAH vs. control group, $p < 0.05$.

Furthermore, we compared the TMAD indices between the two PAH groups. Although all were lower in the worsening group compared to the non-worsening group, only the TMADm% was significantly decreased in worsening PAH children ($p = 0.010$) (Table 2, Figure 5).

We have performed ROC curve analysis for TMADm% and BNP level that showed significant difference between non-worsening and worsening PAH groups. We found a cut-off value of 16.15% for TMADm% that can predict worsening in PAH children, with a sensitivity and specificity of 88.90% and 72.70%, respectively (Figure 6).

Additionally, we found a cut-off value of 34.35 pg/mL for BNP level that can predict worsening in PAH children, with a sensitivity and specificity of 77.80% and 72.70%, respectively (Figure 7).

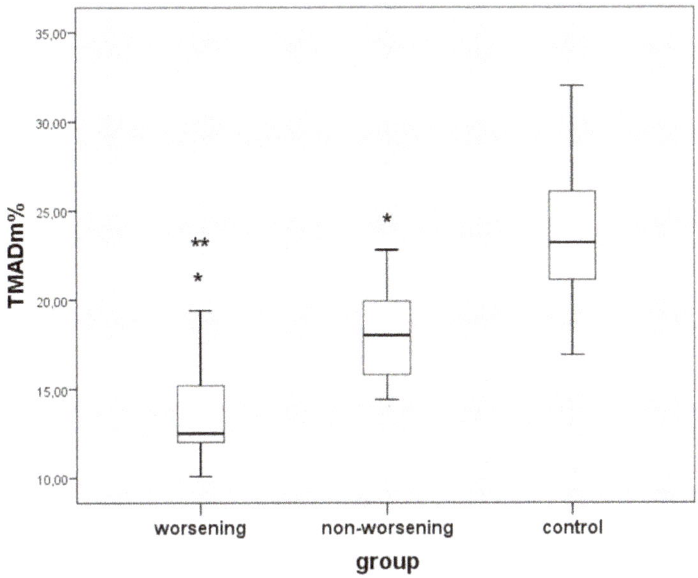

Figure 5. Comparison between worsening, non-worsening PAH children and controls regarding TMADm%; * non-worsening or worsening PAH vs. controls; ** worsening vs. non-worsening PAH group; $p < 0.05$.

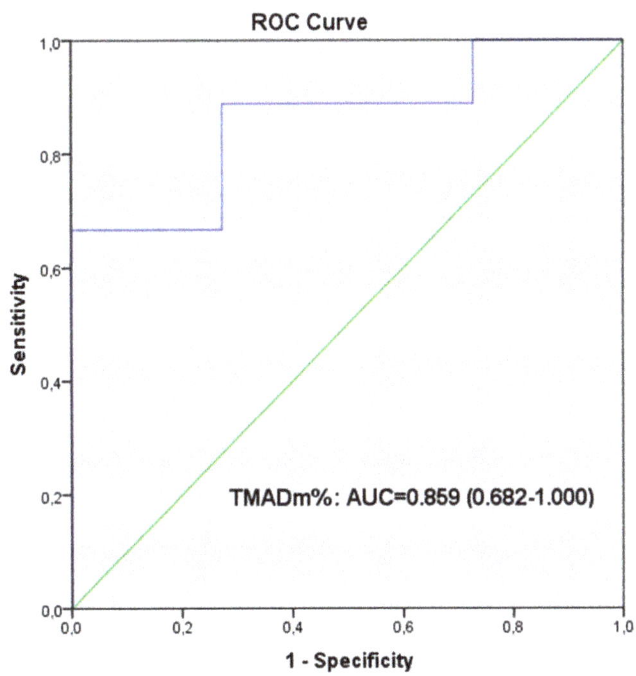

Figure 6. Receiver-operator characteristic analysis. ROC curves result from TMADm% that showed significant difference between non-worsening and worsening PAH groups. The optimum cut-off value for TMADm% to predict clinical worsening was 16.15%; the sensitivity and specificity were 88.90% and 72.70%, respectively.

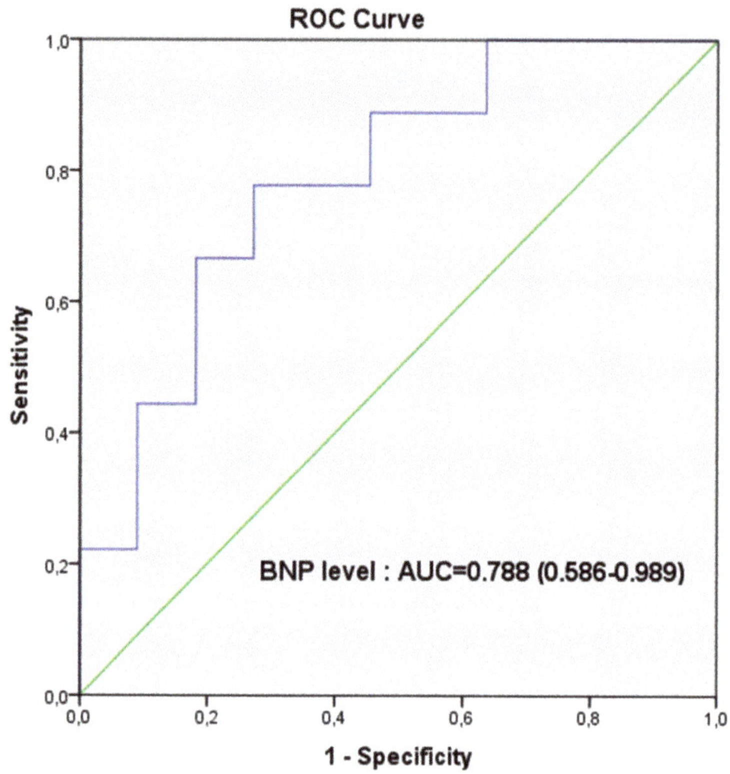

Figure 7. Receiver-operator characteristic analysis. ROC curves result from BNP level that showed significant difference between non-worsening and worsening PAH groups. The optimum cut-off value for BNP level to predict clinical worsening was 34.35 pg/mL; the sensitivity and specificity were 77.80% and 72.70%, respectively.

4. Discussion. This Study Emphasizes the Prognostic Value of TMAD of the Tricuspid Valve in PAH Children

4.1. Parameters for Assessing RV Function

Evaluation of the right ventricle is paramount in PAH patients, considering that right ventricular function determines the prognosis and the treatment responsiveness in these patients [16].

Some conventional echocardiographic parameters such as RA dimension, RV dimension, TAPSE, S/D ratio, FAC, pulmonary artery acceleration time, have been suggested as predictors of outcome in PAH patients (Nice, 2018) [3,4]; however, the need for identification of more valid, easy-to-determine predictors and treatment goals in pediatric patients with PAH is emphasized [3].

In our study, we found a significant decrease in the following conventional parameters RV-FAC, TAPSE, S′ or increase in the following conventional parameters RV-MPI, RV S/D ratio, LV-EI between non-worsening PAH children compared to controls and between worsening PAH children compared to controls. In addition, these parameters were more affected in the worsening group in comparison with the non-worsening group, however, in our study, there was no significant difference between the two groups. We speculate that this could be explained by the fact that in our study, the majority of patients in the worsening group has a baseline WHO functional class ≤ 2, so in that phase of the disease, the conventional echocardiographic parameters are not sensitive enough to predict worsening.

4.2. RV Longitudinal Function and TMAD of the Tricuspid Valve

Anatomical studies have shown that the longitudinal fibers are predominant in the RV, localized in the subendocardial layer, whereas less circumferential fibers are found in the superficial layer in healthy children. This explains why the longitudinal shortening is a greater contributor to RV stroke volume [17–19].

Echocardiographic assessment of RV longitudinal function, according to the present guidelines [3,13,14], includes the following parameters: FAC, TAPSE and RV strain measurements.

The TMAD is a novel parameter derived from speckle-tracking echocardiography and is based on tracking bright echoes within the myocardium [15]. Although it was determined offline in our study with specialized software, the new echo machines allow an easy way to obtain this parameter.

The TMAD of tricuspid valve is a method to evaluate the longitudinal motion of the tricuspid annulus in different points (lateral, medial and midpoint). It was validated against magnetic resonance imaging (MRI) derived RV ejection fraction (EF) in assessing RV function [10,20]. Although tricuspid annular plane systolic excursion (TAPSE) using M-mode is a widely used parameter for assessing RV function, it evaluates only the motion of the tricuspid annulus. We speculate that TMAD, especially TMADm%, are more reliable parameters as they consider the complete longitudinal contraction (from base to apex), so is factoring the RV length from end-diastole to end-systole.

Our study demonstrated that TMADlat was higher than TMADmid and TMADsept, so the values differed based on the site of measurements. Li, Y. et al. found the same distribution of the different TMAD values [20]. According to these values, the displacement was higher at the level of the RV free wall than the septal wall, likely explained by the interventricular mechanical coupling between the two ventricles through the interventricular septum. This finding is consistent with prior studies demonstrating the ventricular interdependence [20,21].

Moreover, tricuspid annulus midpoint fractional displacement (TMADm%) also known as right ventricular longitudinal shortening fraction (RV-LSF), calculated as the maximum displacement of the tricuspid midpoint throughout the cardiac cycle, is a more complex parameter that factors the RV length throughout the cardiac cycle. The TMADm% has been studied as a marker of the RV dysfunction. Maniwa et al. found that a poor RV systolic function, expressed by 3-dimensional echocardiography—measured RVEF < 45%—was detected by a TMADm% < 14.7% with a high sensitivity and specificity of 93% and 95%, respectively, followed by RV free wall longitudinal strain [22]. Li et al. found that TMADm% is the best predictor for RV dysfunction expressed by MRI-derived RVEF in PAH patients, with an optimal cut-off value of 16.05% [20]. In a study conducted in infants (age ranged from 3 to 11 months) with dominant right ventricular physiology who underwent bidirectional Glenn anastomosis, Penk et al. demonstrated that TMAD of the tricuspid valve is an independent predictor of midterm mortality or heart transplantation in these children, with a mean value of 10% for non-survivors or heart transplanted [23]. Beyls et al. found that TMADm% (named RV longitudinal shortening fraction in their study) was markedly decreased in COVID-19 patients with acute cor pulmonale, and a cut-off value of 17% is able to identify RV dysfunction in these patients, with high specificity and sensibility [24].

4.3. Prognostic Value of TMAD in PAH Children

The prognostic value of TMAD of the tricuspid valve in PAH children was not studied so far. We found that, although all TMAD parameters are lower in worsening group as compared to non-worsening group, only TMADm% is significantly lower ($p < 0.05$). Furthermore, we showed that a cut-off value of 16.15% for TMADm% is associated with clinical worsening in PAH children.

In addition to TMADm%, BNP level is significantly lower in worsening group as compared to non-worsening group ($p < 0.05$). However, comparing the ROC curves for the two parameters (TMADm% and BNP level), we observed a better sensitivity of TMADm%

to predict worsening, expressed by the area under curve: AUC = 0.859 (0.682–1.000) for the TMADm% in comparison to AUC = 0.788 (0.586–0.989) for the BNP level.

According to our knowledge, this is the first study that describes the predictive value of TMAD in children with PAH.

5. Conclusions

Tricuspid annulus midpoint fractional displacement, namely TMADm%, an angle-dependent speckle-tracking derived parameter, could be a good additional parameter in the assessment of the longitudinal RV systolic function. Since TMADm% is significantly lower in worsening PAH group, it can be considered a good parameter in predicting clinical worsening in children with PAH.

6. Study Limitations

The first limitation of our study is that we have not compared TMAD with MRI which is currently the gold standard for RV ejection fraction assessment. However, several studies have already demonstrated good correlation between TMAD of the tricuspide valve measured by 2D-speckle tracking and MRI-derived RV ejection fraction [10,20,21]. The small number of cases included in the present study is another limitation, so we consider that our findings need to be confirmed by further studies with a larger number of cases.

Author Contributions: Conceptualization, I.M., C.C.Ș. and R.T.; Methodology, I.M., C.C.Ș. and M.M.; Formal Analysis, I.M.; Investigation, I.M., M.M. and C.C.Ș.; Resources, I.M., A.F., M.M. and R.T.; Data Curation, I.M.; Writing—Original Draft Preparation, I.M., C.C.Ș. and A.F.; Writing—Review & Editing, I.M., A.F. and R.T.; Supervision, R.T.; Project Administration, I.M. and A.F. All authors have read and agreed to the published version of the manuscript.

Funding: This work was supported by the University of Medicine, Pharmacy, Science and Technology "George Emil Palade" of Târgu Mureș Research Grant number 511/5/17 January 2022.

Institutional Review Board Statement: The study was conducted according to the guidelines of the Declaration of Helsinki and approved by the Ethics Committee of University of Medicine, Pharmacy, Science and Technology "George Emil Palade" of Târgu Mureș (Approval Number: 1596/2 February 2022).

Informed Consent Statement: Written informed consent was obtained from all the patients' parents/legal guardian included in the study.

Data Availability Statement: The data presented in this study are available on request from the corresponding author.

Conflicts of Interest: The authors declare no conflict of interest.

References

1. Simonneau, G.; Montani, D.; Celermajer, D.S.; Denton, C.P.; Gatzoulis, M.A.; Krowka, M.; Williams, P.G.; Souza, R. Haemodynamic definitions and updated clinical classification of pulmonary hypertension. *Eur. Respir. J.* **2019**, *53*, 1801913. [CrossRef] [PubMed]
2. Koestenberger, M.; Apitz, C.; Abdul-Khaliq, H.; Hansmann, G. Transthoracic echocardiography for the evaluation of children and adolescents with suspected or confirmed pulmonary hypertension. Expert consensus statement on the diagnosis and treatment of paediatric pulmonary hypertension. The European Paediatric Pulmonary Vascular Disease Network, endorsed by ISHLT and DGPK. *Heart* **2016**, *102* (Suppl. 2), ii14–ii22. [PubMed]
3. Hansmann, G.; Koestenberger, M.; Alastalo, T.P.; Apitz, C.; Austin, E.D.; Bonnet, D.; Budts, W.; D'Alto, M.; Gatzoulis, M.A.; Hasan, B.S.; et al. 2019 updated consensus statement on the diagnosis and treatment of pediatric pulmonary hypertension: The European Pediatric Pulmonary Vascular Disease Network (EPPVDN), endorsed by AEPC, ESPR and ISHLT. *J. Heart Lung Transplant.* **2019**, *38*, 879–901. [CrossRef] [PubMed]
4. Rosenzweig, E.B.; Abman, S.H.; Adatia, I.; Beghetti, M.; Bonnet, D.; Haworth, S.; Ivy, D.D.; Berger, R.M. Paediatric pulmonary arterial hypertension: Updates on definition, classification, diagnostics and management. *Eur. Respir. J.* **2019**, *53*, 1801916. [CrossRef] [PubMed]
5. Asada, D.; Okumura, K.; Ikeda, K.; Itoi, T. Tissue motion annular displacement of the mitral valve can be a useful index for the evaluation of left ventricular systolic function by echocardiography in normal children. *Pediatr. Cardiol.* **2018**, *39*, 976–982. [CrossRef] [PubMed]

6. Black, D.E.; Bryant, J.; Peebles, C.; Godfrey, K.M.; Hanson, M.; Vettukattil, J.J. Tissue motion annular displacement of the mitral valve using two-dimensional speckle tracking echocardiography predicts the left ventricular ejection fraction in normal children. *Cardiol. Young* **2014**, *24*, 640–648. [CrossRef]
7. Liu, L.; Tuo, S.; Zhang, J.; Zuo, L.; Liu, F.; Hao, L.; Sun, Y.; Yang, L.; Shao, H.; Qi, W.; et al. Reduction of left ventricular longitudinal global and segmental systolic functions in patients with hypertrophic cardiomyopathy: Study of two-dimensional tissue motion annular displacement. *Exp. Ther. Med.* **2014**, *7*, 1457–1464. [CrossRef]
8. Liu, S.; Ren, W.; Zhang, J.; Ma, C.; Yang, J.; Zhang, Y.; Guan, Z. Incremental value of the tissue motion of annular displacement derived from speckle-tracking echocardiographu for differentiating chronic constrictive pericarditis from restrictive cardiomyoptahy. *J. Ultrasound Med.* **2018**, *37*, 2637–2645. [CrossRef]
9. Alonso, P.; Andres, A.; Miro, V.; Igual, B.; Sánchez, I.; Salvador, A. Diagnostic power of echocardiographic speckle tracking of the tricuspid annular motion to assess right ventricular dysfunction. *Int. J. Cardiol.* **2014**, *172*, e218–e219. [CrossRef]
10. Ahmad, H.; Mor-Avi, V.; Lang, R.M.; Nesser, H.J.; Weinert, L.; Tsang, W.; Steringer-Mascherbauer, R.; Niel, J.; Salgo, I.S.; Sugeng, L. Assessment of right ventricular function using echocardiographic speckle tracking of the tricuspid annular motion: Comparison with cardiac magnetic resonance. *Echocardiography* **2012**, *29*, 19–24. [CrossRef]
11. Yang, H.S.; Kim, T.Y. Intraoperative transesophageal echocardiography in tricuspid valve surgery. *Clin. Ultrasound* **2019**, *4*, 67–77. [CrossRef]
12. Muntean, I.; Benedek, T.; Melinte, M.; Suteu, C.; Toganel, R. Deformation pattern and predictive value of right ventricular longitudinal strain in children with pulmonary arterial hypertension. *Cardiovasc. Ultrasound* **2016**, *14*, 27. [CrossRef] [PubMed]
13. Lopez, L.; Colan, S.D.; Frommelt, P.C.; Ensing, G.J.; Kendall, K.; Younoszai, A.K.; Lai, W.W.; Geva, T. Recommendations for Quantification Methods During the Performance of a Pediatric Echocardiogram: A Report from the Pediatric Measurements Writing Group of the American Society of Echocardiography Pediatric and Congenital Heart Disease Council. *J. Am. Soc. Echocardiogr.* **2010**, *23*, 465–495. [CrossRef] [PubMed]
14. Rudski, L.G.; Lai, W.W.; Afilalo, J.; Hua, L.; Handschumacher, M.D.; Chandrasekaran, K.; Solomon, S.D.; Louie, E.K.; Schiller, N.B. Guidelines for the echocardiographic assessment of the right heart in adults: A report from the American Society of Echocardiography endorsed by the European Association of Echocardiography, a registered branch of the European Society of Cardiology, and the Canadian Society of Echocardiography. *J. Am. Soc. Echocardiogr.* **2010**, *23*, 685–713. [PubMed]
15. Geyer, H.; Caracciolo, G.; Abe, H.; Wilansky, S.; Carerj, S.; Gentile, F.; Nesser, H.J.; Khandheria, B.; Narula, J.; Sengupta, P.P. Assessment of myocardial mechanics using speckle tracking echocardiography: Fundamentals and clinical applications. *J. Am. Soc. Echocardiogr.* **2010**, *23*, 351–369. [CrossRef]
16. Condon, D.F.; Nickel, N.P.; Anderson, R.; Mirza, S.; de Jesus Perez, V.A. The 6th World Symposium on Pulmonary Hypertension: What's old is new. *F1000Research* **2019**, *8*, 888. [CrossRef]
17. Ho, S.H.; Nihoyannopoulos, P. Anatomy, echocardiography and normal right ventricular dimensions. *Heart* **2006**, *92* (Suppl. 1), i2–i13. [CrossRef]
18. Carlsson, M.; Ugander, M.; Heiberg, E.; Arheden, H. The quantitative relationship between longitudinal and radial function in left, right, and total heart pumping in humans. *Am. J. Physiol. Heart Circ. Physiol.* **2007**, *293*, H636–H644. [CrossRef]
19. Rushmer, R.F.; Crystal, D.K.; Wagner, C. The functional anatomy of ventricular contraction. *Circ. Res.* **1953**, *1*, 162–170. [CrossRef]
20. Li, Y.; Wang, Y.; Yang, Y.; Liu, M.; Meng, X.; Shi, Y.; Zhu, W.; Lu, X. Tricuspid annular displacement measured by 2-dimensional speckle tracking echocardiography for predicting right ventricular function in pulmonary hypertension. *Medicine* **2018**, *97*, e11710. [CrossRef]
21. Li, Y.D.; Wang, Y.D.; Zhai, Z.G.; Guo, X.J.; Wu, Y.F.; Yang, Y.H.; Lu, X.Z. Relationship between echocardiographic and cardiac magnetic resonance imaging-derived measures of right ventricular function in patients with chronic thromboembolic pulmonary hypertension. *Thromb. Res.* **2015**, *135*, 602–606. [CrossRef] [PubMed]
22. Maniwa, N.; Hozumi, T.; Takemoto, K.; Wada, T.; Kashiwagi, M.; Shimamura, K.; Shiono, Y.; Kuroi, A.; Matsuo, Y.; Ino, Y.; et al. Value of tissue-tracking tricuspid annular plane by speckle-tracking echiocardiography for the assessment of right ventricular systolic dysfunction. *Echocardiography* **2018**, *36*, 110–118. [CrossRef] [PubMed]
23. Penk, J.S.; Zaidi, S.J.H.; Lefaiver, C.A.; Muangmingsuk, S.; Cui, V.W.; Roberson, D.A. Tissue motion annulat displacement predicts mortality/transplant after the bidirectional Glenn. *World J. Pediatr. Congenit. Heart Surg.* **2018**, *9*, 171–176. [CrossRef] [PubMed]
24. Beyls, C.; Bohbot, Y.; Huette, P.; Booz, T.; Daumin, C.; Abou-Arab, O.; Mahjoub, Y. Usefulness of Right Ventricular Longitudinal Shortening Fraction to Detect Right Ventricular Dysfunction in Acute Cor Pulmonale Related to COVID-19. *J. Cardiothorac. Vasc. Anesth.* **2021**, *35*, 3594–3603. [CrossRef] [PubMed]

applied sciences

MDPI

Article

Automated Classification of Left Ventricular Hypertrophy on Cardiac MRI

Adam Budai [1,*], Ferenc Imre Suhai [2], Kristof Csorba [1], Zsofia Dohy [2], Liliana Szabo [2], Bela Merkely [2] and Hajnalka Vago [2]

1 Department of Automation and Applied Informatics, Faculty of Electrical Engineering and Informatics, Budapest University of Technology and Economics, H-1111 Budapest, Hungary; kristof@aut.bme.hu
2 Heart and Vascular Center, Semmelweis University, H-1122 Budapest, Hungary; suhai.ferenc.imre@gmail.com (F.I.S.); dohyzsofi@gmail.com (Z.D.); sz.liliana.e@gmail.com (L.S.); merkely.bela@gmail.com (B.M.); vagoha@gmail.com (H.V.)
* Correspondence: budai.adam@aut.bme.hu

Abstract: Left ventricular hypertrophy is an independent predictor of coronary artery disease, stroke, and heart failure. Our aim was to detect LVH cardiac magnetic resonance (CMR) scans with automatic methods. We developed an ensemble model based on a three-dimensional version of ResNet. The input of the network included short-axis and long-axis images. We also introduced a standardization methodology to unify the input images for noise reduction. The output of the network is the decision whether the patient has hypertrophy or not. We included 428 patients (mean age: 49 ± 18 years, 262 males) with LVH (346 hypertrophic cardiomyopathy, 45 cardiac amyloidosis, 11 Anderson–Fabry disease, 16 endomyocardial fibrosis, 10 aortic stenosis). Our control group consisted of 234 healthy subjects (mean age: 35 ± 15 years; 126 males) without any known cardiovascular diseases. The developed machine-learning-based model achieved a 92% F1-score and 97% recall on the hold-out dataset, which is comparable to the medical experts. Experiments showed that the standardization method was able to significantly boost the performance of the algorithm. The algorithm could improve the diagnostic accuracy, and it could open a new door to AI applications in CMR.

Keywords: classification; left ventricular hypertrophy; CMR; machine learning

Citation: Budai, A.; Suhai, F.I.; Csorba, K.; Dohy, Z.; Szabo, L.; Merkely, B.; Vago, H. Automated Classification of Left Ventricular Hypertrophy on Cardiac MRI. *Appl. Sci.* **2022**, *12*, 4151. https://doi.org/10.3390/app12094151

Academic Editor: Julio Garcia Flores

Received: 16 March 2022
Accepted: 16 April 2022
Published: 20 April 2022

1. Introduction

Cardiovascular diseases are the leading cause of death in developed countries [1,2]. Cardiovascular magnetic resonance (CMR) provides functional and morphological information of the heart for the evaluation, management, and diagnosis of patients with suspected or established cardiovascular disease. CMR is a multi-parametric, non-invasive imaging modality, which is considered the gold standard for the assessment of global and regional function and is able to evaluate myocardial perfusion and viability, tissue characterization, and coronary artery anatomy [3]. Left ventricular hypertrophy (LVH) is present in 15% to 20% of the population. It is more common in Afro-Americans and in patients with hypertension and obesity [4]. LVH is an independent predictor of future cardiovascular events, including coronary heart disease, heart failure, and stroke, regardless of its etiology [5,6]. The definition of LVH is an increase in left ventricular mass either due to an increase in wall thickness, an increase in cavity size, or both. In clinical practice, LVH is a common condition, which can be caused by diverse physiological and pathological mechanisms such as athlete's heart, hypertension, aortic stenosis, hypertrophic cardiomyopathy, infiltrative heart muscle disease, storage, and metabolic disorders (amyloidosis, Anderson–Fabry disease, etc.). LVH can develop silently over several years without symptoms, and it can be difficult to diagnose. The electrocardiogram (ECG) is a useful, but less sensitive tool for detecting LVH. The utility of the ECG lies in its relative inexpensiveness and wide availability. Its limitations stem from its moderate sensitivity or specificity, depending on which of the various diagnostic criteria are applied [7,8]. In the Multi-Ethnic Study of Atherosclerosis patients who underwent MRI and ECG, it was found that various ECG criteria had low sensitivity for the detection of LVH [9]. As a result of these limitations of ECG, LVH is most reliably identified on imaging with echocardiography or CMR. Prior studies primarily used ECG [10] or M-Mode and two-dimensional (2D) echocardiography to identify LVH.

Conventional 2D echocardiography is the first-line imaging modality, which is used to evaluate the patterns, extent, and distribution of LVH and other anatomic and functional parameters and ventricular function. Nonetheless, echocardiography is limited by intra-observer and inter-observer variability, acoustic windows, and the lack of tissue characterization. Echocardiography-based LVH evaluation varies among the different definitions by ultrasound technicians and laboratories around the world, leading to inconsistency among epidemiological studies, and therefore, this could limit its clinical application [11]. CMR provides a comprehensive evaluation of myocardial hypertrophy regarding the extent and distribution of LVH and tissue characterization. Accurate measurements of wall thickness, the phenotype of hypertrophy, chamber size, and ventricular function can be obtained without any limiting factors such as imaging windows and body habitus. Importantly, CMR has a myocardial tissue characterization property that allows phenotypic determination of the LVH and careful evaluation of the precise etiology of LVH, which is a challenging clinical problem [12]. Hypertrophic cardiomyopathy (HCM) is the most common inherited cardiac disease, which often leads to sudden death in young people with an estimated prevalence of about 1:500 [13]. LVH is characterized by sudden cardiac death, stroke, and heart failure, but also decreased life expectancy [14,15]. The different LVH morphologic pattern can be precisely assessed by CMR and able to identify segmental hypertrophy, which can be difficult for echocardiography (i.e., apical HCM).

The application of machine learning methods on CMR images has boomed in the past 5–6 years. A huge body of work is available on automatic processing of MRI images, such as the segmentation of the ventricles [16–18], left ventricular quantification [19], pediatric cardiomyopathy classification [20], left ventricle wall motion classification [21], and cardiovascular event prediction for dilated cardiomyopathy [22]. These research endeavors also resulted in new architectures, developed for this field specifically, e.g., U-net [23] for myocardium segmentation, ν-net for cardiac vascular segmentation [24], and Ω-net for multiview CMR detection, orientation, and segmentation [25]. Besides the algorithmic improvement, the increasing availability of public datasets fuels the breakthroughs in the healthcare domain [26,27]. For LVH, the lack of publicly available benchmarks is reflected in the lower number of papers on the subject [28]. However, for echocardiography, there are promising works out there [29–31]. For instance, in paper [32], the echocardiography-based hypertrophy detection calculates the wall thickness, then the decision is made. This method automates the wall thickness measurements with thresholding, then the wall thickness is calculated from the adjusted contours. The accuracy of echocardiography-based diagnosis tends to be lower than that of CMR-based examinations [33,34]. This further motivates the application of CMR for hypertrophy detection. In the work of [35], the disease classification was based on a multi-stage process. First, it segments the heart, then it calculates the volumetric data. The volumetric data are used as features in a random forest to obtain the classification. The classes are different from ours, but there are HCM and normal heart cases, while the input images are CMR images. In paper [36], the authors developed an automatic wall thickness measurement on CMR images. The measurement was based on endocardial and epicardial segmentations. Another method relies on the clinical assessment of normal ranges for different morphological characteristics [37]. To the best of our knowledge, this is the first paper discussing an automatic method for left ventricular hypertrophy classification from CMR images.

If the algorithm can detect suspicious features of hypertrophy during a regular CMR examination, this indicates that the applied CMR protocol should be changed by adding necessary measurements and sequences on-site for a more detailed evaluation without the need for additional examinations. During the post-process evaluation, it could improve the diagnostic accuracy by recognizing a milder, incipient form of LVH, which can be challenging for the less-experienced readers. The early detection of LVH and appropriate therapy will decrease cardiovascular morbidity and mortality [38]. In this paper, steps toward this ambition were made by developing an algorithm that considers more views of the heart and classifies the patient's hearts as normal or exhibiting hypertrophy. The algorithm we developed achieved results comparable to the human readers. Its high recall and sufficient precision allow for its use in an on-site setting, potentially causing the operators to change the CMR protocol (e.g., to administer the contrast agent, acquire late enhancement images, etc.) if hypertrophy is suspected. During the CMR examination, usually, the long-axis cine images are acquired first, then the short-axis cine images, then the late enhancement images if needed. We found that if the algorithm is restricted to only use long-axis cine images, it is still sufficient to alert the operator in order to select an appropriate CMR protocol, but might be limited in some selected cases. The rest of the paper is structured the following way: In Section 2, we introduce the dataset we utilized during our research, then we describe how our method works. In Section 3, we report the experimental results on a hold-out dataset and we make a comparison to the human-level performance. Section 4 describes our concluding thoughts.

2. Materials and Methods

The goal of this research is to develop an algorithm for hypertrophy classification from CMR scans. The scans contain more views: short-axis, long-axis. Our dataset was collected from the database of the The Heart and Vascular Center of Semmelweis University. Our method is based on the raw image scans with all available views, and the classification result is the direct output; we did not calculate intermediate features such as wall thickness.

2.1. Dataset

After the exclusion of patients with poor image quality, we investigated 428 patients (mean age: 49 $\pm$ 18 years, 262 males) with left ventricular hypertrophy in whom CMR examination was clinically indicated and 234 healthy subjects (mean age: 35 $\pm$ 15 years; 126 males) without any known cardiovascular diseases as a control group. The patients underwent CMR examination in our tertiary referral center between January 2009 and February 2019. Out of the 428 LVH patients, 346 had HCM (age: 46.9 $\pm$ 18.2 144 males), 45 patients had cardiac amyloidosis (age: 63.9 $\pm$ 9.7 years, 26 males), 11 patients had Anderson–Fabry disease (age: 48.3 $\pm$ 12.9 years, 7 males), 16 patients had endomyocardial fibrosis (age: 46.4 $\pm$ 14.3, years 9 males), and 10 patients had aortic stenosis (age: 63.4 $\pm$ 17.5 years, 5 males). Appendix C shows example images. CMR examinations were performed on a 1.5 T magnetic resonance (MR) scanner (Achieva, Philips Medical Systems) using a cardiac coil. ECG gated balanced steady-state free precession (bSSFP) cine images were acquired in the three standard long-axis views: 2-chamber, 4-chamber, and LV outflow tract views. The protocol used for cine images in the present study was described in detail in a previous publication [39]. Short-axis (SA) images were also acquired with the full coverage of the left ventricle.

2.2. Model Architecture

The algorithm decides whether the patient has hypertrophy. The input to the algorithm is created from CMR scans of 4 views (axis). We used multi-view data because hypertrophy classification is challenging and different views provide different information. It is possible to see a pattern on a short-axis image that cannot be seen on the long-axis images or the other way around. The input images were collected from 4 views:

- Short-axis images from the apex to the base at different stages of the cardiac cycle;
- Long-axis, two-chamber images at different stages of the cardiac cycle (heart beat);
- Long-axis, three-chamber images at different stages of the cardiac cycle;
- Long-axis, four-chamber images at different stages of the cardiac cycle.

The usage of all the images from the short-axis scan has difficulties. The input would be too big, and the number of images were not the same for all patients. Therefore, for the short-axis view, we took three images in each second phase of the cardiac cycle. At each chosen cardiac cycle, we used one image from the basal, one from the mid, and one from the apical region, resulting in 36 images; see Figures 1 and 2. In the case of the long-axis views, we took each second image from the cardiac cycle, resulting in 12 images; see Figure 3.

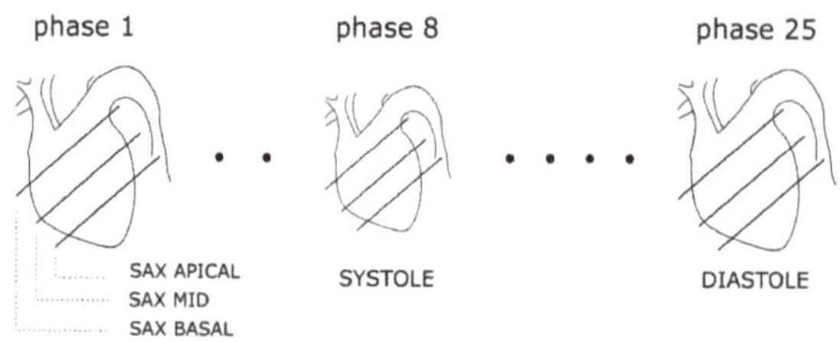

Figure 1. The illustration of how the short-axis images are created. The systole is the phase where the heart volume is the lowest, while in the diastole, it is the highest. A heart beat (cardiac cycle) is divided into 25 phases. In each phase, several slices are scanned. See the parallel lines. There are three major regions for the heart: basal, mid, apical.

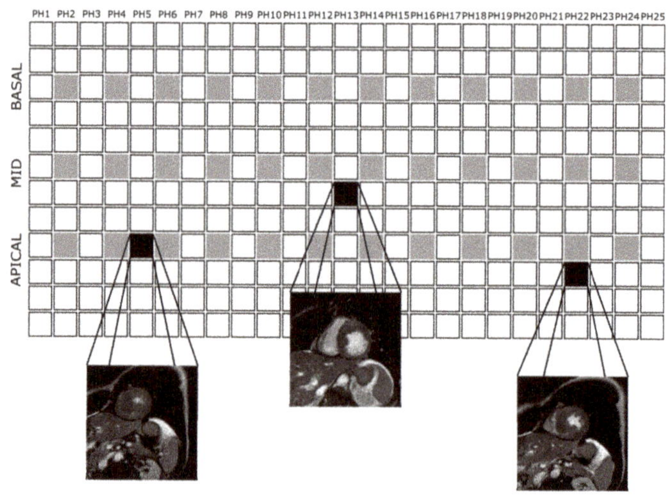

Figure 2. A short-axis CMR scan produces around 400 images for a heart. There are 12–16 slices visited by the scanner at each phase (time point). During one heart beat (cardiac cycle), 25 images are created for a slice. There are three main regions: apical, mid, basal. See also Figure 1. To create a fixed-size input for the model, we picked three images from each second phase; see the gray boxes. Overall, this results in 36 images. Black squares show example images, how the real image looks at a given slice and phase.

The model is an ensemble of the extractors of the separate views. Images from each view are fed into a separate network to extract features. The features are concatenated, then the ensemble classifier is applied to obtain the prediction (normal or exhibiting hypertrophy); see Figure 4. The architecture of the extractor for the best-performing model can be seen in Figure 4, left side. We used the same extractor for each view. The extractors were trained separately; therefore, a temporary layer was applied to create a temporary classifier. The architecture of the temporary layer can be seen in Table 1. After the extractors are trained, an ensemble model is created with an ensemble classifier; see Figure 4, right side, under the classifier block. The models were built from residual blocks, with each block containing 3-dimensional convolutions and batch normalizations; see Figure 4, bottom part.

The reason for the 3D convolution is the positive effect of considering the time dimension of the input (how the heart moves). For further elaboration on the performance and the choices we made, see the details in Section 3.3.

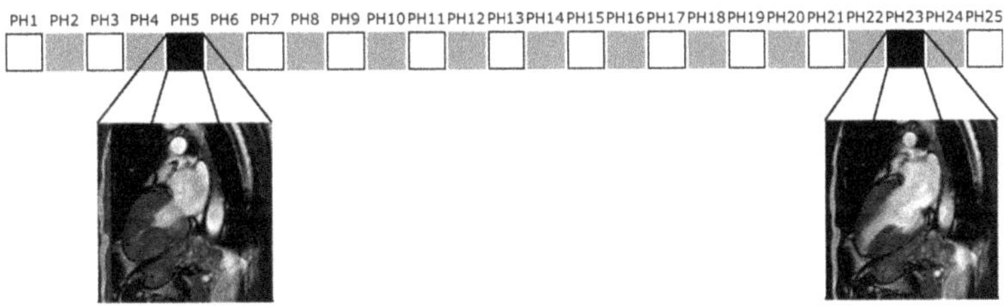

Figure 3. Long-axis CMR scan has three view: 2-chamber, 4-chamber, and 3-chamber views. For each view, 25 images are produced. The images are created from the same slice of the heart, but at different time points of the cardiac cycle. To create a fixed-size input, we picked an image from every second phase; see the gray boxes. Overall, this results in 12 images for each view (LA2, LA4, LA3).

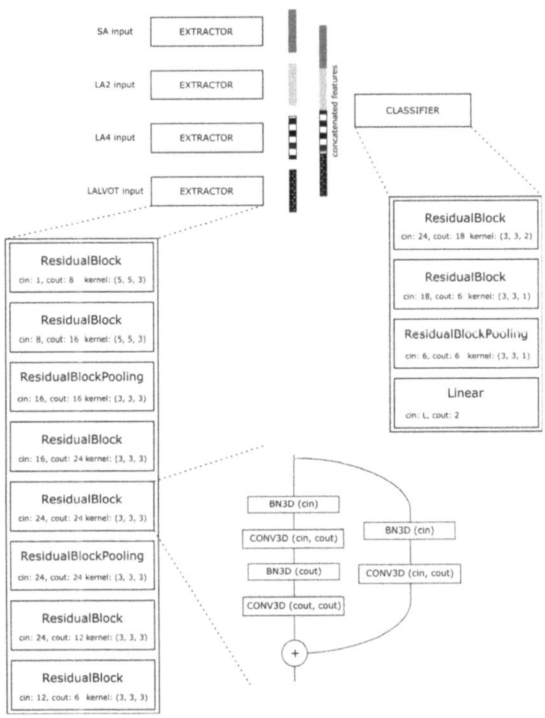

Figure 4. The schematic architecture of the model. Each view has an extractor and the output features from the extractors aggregated by concatenating the features along the channel dimension. This is the ensemble of the views. The extractor and the classifier are built from 3D residual blocks. The architecture of the 3D ResNet blocks can be seen in the middle of the image. The ResidualBlock and the ResBlockPooling differ in the strides. For the pooling block, the first convolution and the convolution on the skip branch have stride 2; otherwise, it is 1. The activations are ReLUs, which were applied after the batch normalization layers. In the ResidualBlock version, we applied padding in each convolution, while in pooling, we applied padding in the last convolution of the straight branch. Padding was: $(k-1)/2$ for each dimension. The kernel sizes were chosen as odd values in each case.

Table 1. The architecture of the temporary classifier. After the pooling layer, the tensors are reshaped to (batch, L) size. The value of L is different for the long-axis and short-axis views.

Block	Block Name	Cin	Cout	Kernel Size
TC1	ResBlockPooling	6	3	(3, 3, 1)
TC2	Linear	L	2	-

2.3. Preprocessing and Data Augmentation

Before the images are fed into the model, two main steps are executed: (1) preprocessing and (2) augmentation. The augmentation is the same for each of the views, but preprocessing contains an additional step for the long-axis views. Preprocessing always applies noise reduction by cropping the intensity values between the 1st and 99th percentiles. Then, the images are normalized into a 0–1 interval. For the long-axis views, the images are standardized because their orientation shows high variance. Standardization is achieved by a superposition to a reference frame calculated for each view separately. The reference frame is given as the normal vector of a typical image for a given view. The superposition applies mirroring and a rotation around the center point of the image to be preprocessed. Appendix A gives further insight into the details of the standardization. In Section 3.3, further details are shown about the effect of the standardization on performance. The augmentation contains a random rotation and Gaussian noise.

2.4. Training Scheme

We trained the model in two stages. This training process falls into the supervised learning paradigm, because we have the ground truth pathologies for each scan. The dataset was unbalanced; therefore, we sampled the normal group with higher probability to equalize the occurrences of hypertrophic and normal samples in the training batches; see Section 2.1 for the ratio. The dataset was split into three parts: training (70%), validation (15%), and testing (15%). The test set was created only once, and we kept it until the final test with the best model chosen on the validation test. We repeated the training with each parameter setting three times to understand the stability of the results. In each repetition, the training and validation parts were resampled. First, the feature extractors were trained separately to predict whether the patient had hypertrophy. For this part, we used a temporary layer at the end of each extractor to create a classifier. Then, the temporary layer was removed, and the ensemble model was built. For combining the outputs of the feature extractors, we concatenated (the long-axis was padded in the depth dimension) the features, then fed them into the classifier; see Figure 4. The whole ensemble was trained, but the feature extractors' weights were frozen. The training was applied on different combinations of the possible views. The combinations were based on realistic scenarios, because the earlier we can detect the condition of hypertrophy, the faster the operators can react during the scanning procedure. In a clinical setting, the examination process mostly follows similar orders among the views. During the CMR examination, the typical order was long-axis views, then short-axis view. It is important to test only using the long-axis view, the short-axis view, and then, their combination. The parameters of the best model can be seen in Table A1.

2.5. Human Evaluation

The performance of the algorithm was also compared to human experts (hearafter readers). The design of the evaluation simulated a realistic setup for an everyday examination procedure. The readers were asked to read CMR scans of 117 subjects, but they were not told the real purpose of the study. About each subject, a very brief patient history was provided (without giving clear reference to the real disease) along with the images of a full MRI scan. This included the short-axis and long-axis images. For the analysis, we included CMR scans from the normal group as well and the following pathologies: acute or chronic myocardial infarction, dilated cardiomyopathy, Takotsubo cardiomyopathy, and acute myocarditis. The list contained the most frequent pathologies encountered during regular assessments. We also included different cardiac pathologies that could cause LVH (HCM, Anderson–Fabry disease, amyloidosis, aortic stenosis, and endomyocardial fibrosis). The reason for pathologies outside of hypertrophy was to avoid bias during the evaluation. Overall, six experts finished the experiment. Two of them were senior colleagues (25 and 10 years of experience) and three of them at the mid-senior level (4–7 years of experience), and one of them was a junior (2 years of experience).

3. Results

We experimentally proved that the algorithm described in Section 2 can achieve comparable performance to human experts.

3.1. Results of Human-Evaluation

The human evaluation established a baseline to raise expectation against the algorithm. Table 2 shows the results. Overall means the accuracy of the diagnosis of each expert for all 117 subjects. This includes all the pathologies. In the Hyp-Norm row, the pathologies are grouped into two groups, normal and hypertrophy, which includes all the LVH etiologies considered earlier in this paper. The prediction of a reader was considered as valid if the predicted pathology fell into the hypertrophy group, but the etiology did not have to be accurate. In the HCM row, we measured the accuracy of differentiating between the patients with HCM and other cardiac disorders, which usually represents LVH. In the last three rows, precision, recall, and F1-score were calculated for the Hyp-Norm case. Hypertrophy was considered as a positive event in the confusion matrix. If we compare the consistency among the experts in terms of three groups: normal, hypertrophy, and the rest, we found 83 %, 71 %, and 91 % consistency values, respectively. Consistency is defined as an agreement among at least five radiologists. The high value of recall and the lower value of consistency for the normal group indicates that radiologists tend to classify healthy patients as those having a condition. This is understandable, as a false positive can easily prove to be negative after some further examinations. On the contrary, false negatives can lead to delayed and inappropriate patient care.

Table 2. The scores of the human evaluation. The scores are consistent across the readers; the variance is small.

Scores	R1	R2	R3	R4	R5	R6	Mean
Overall	89.7	91.5	91.5	88.0	89.7	90.6	90.0
Hyp-Norm	92.8	95.7	91.3	94.2	89.9	94.2	93.0
HCM	85.4	89.6	91.7	85.4	93.8	85.4	88.6
Precision	97.8	97.9	88.9	92.3	87.3	95.8	93.3
Recall	93.6	95.8	100	100	100	97.9	97.9
F1	95.7	96.8	94.1	96.0	93.2	96.8	95.4

3.2. Performance of the Algorithm

The performance of the best model can be seen in Table 3. The table shows that only using the LA views was enough to achieve comparable results to humans by considering the standard deviations as well (3–4%). This is important, because the contrast agent can be injected after the long-axis measurements (if the algorithm indicates it and the experts accept it), then the short-axis cine images can be acquired, since the late enhancement images could be acquired at least 10 min after contrast material administration. This approach can save significant amounts of time and can also warn the on-site medical staff that the MRI protocol should be changed in order to avoid further, unnecessary examinations.

Table 3. The performance of the best model on the test sets. The last row shows the result on the validation set when the long-axis views were combined with the short-axis views.

Cases	F1	Precision	Recall
only LA2	90	86	92
only LA4	86	81	90
only LA3	91	86	92
only SA	86	90	83
all LAs	89	84	90
LA+SA	91	88	96
v. LA+SA	93	91	94

The box plots in Figures 5 and 6 were calculated by repeating the test evaluation on 20 randomly sampled subsets of the test data, and in each sample, we used 70% of the test data. This method is similar to bootstrapping. Both images show the same relative performance. The algorithm using only the LA views had lower performance, but when short-axis and long-axis views were combined, the human level and the algorithm scores became close to each other, especially in the case of the recall. The results showed lower F1 and recall for the only short-axis case (see Table 3), which can be a result of the higher complexity of the data. More samples for the SA case could scale up the performance. Similarly, the algorithm (SA+LA) had lower performance than the experts, but we claim that a larger dataset would reduce the gap.

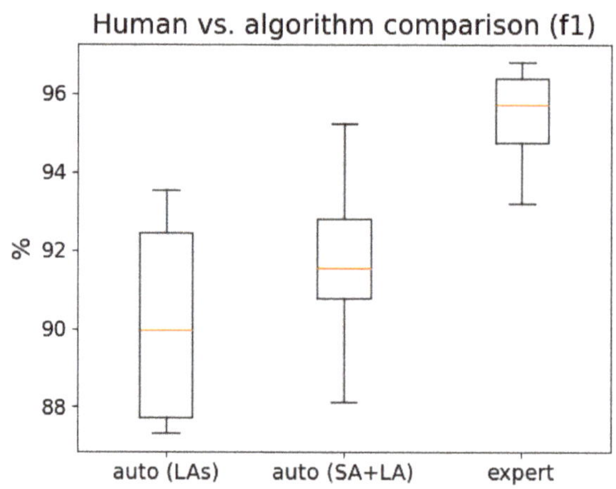

Figure 5. Comparison of the human (expert) and algorithm (auto) performances. The p-value between auto (LA) and auto (SA+LA) is lower than 0.001, which means using the short-axis images contributed to a significantly better performance. Between the auto (SA+LA) and expert group, the p-value was less than 0.001. For calculating the p-values, we used two-sample t-tests.

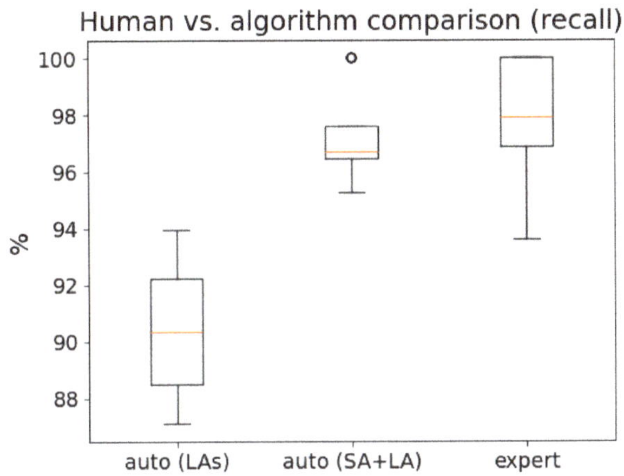

Figure 6. Comparison of the human (expert) and algorithm (auto) performances. High recall is beneficial because the algorithm can identify samples suspicious of hypertrophy with a high probability. The false positives can be handled by the experts who supervise the examination. The *p*-value is less than 0.001 between the auto (LA) and auto (SA+LA) groups. When comparing auto (SA+LA) and the expert groups, we obtained a *p*-value = 0.3, indicating there was no statistically significant difference between them. Therefore, the auto (SA+LA) was statistically identical to the expert group in terms of the recall.

3.3. Ablation Study

We executed several experiments before we arrived at the final model, data processing, and parameter choices. In this subsection, we briefly summarize our findings. We cover the three main aspects of the algorithm:

1. Model selection;
2. Data preprocessing;
3. Hyper-parameter setting.

The above order does not represent the order of our experiments. It was established in order to explain our experience in more logical fashion. We did not measure every possible combination of choices; therefore, we can explain and showcase the tendencies of the different choices.

Model selection. We tried three main architectures. The first architecture was a fully convolutional model with 4–5 convolutional layers, assuming the ensemble model with more views can achieve good results overall and we would not need strong learners per view. Our results indicated that bigger networks would be required to achieve scores (accuracy, F1-score, etc.) around 90 percent. The second architecture was similar to ResNet with two-dimensional convolutions. The time dimension in the long-axis view was stacked together to form a 12-channel image. The structure was similar to the ResNet described in Section 2.2. We experienced significant performance growth (around 3–4 percent) as the model size achieved 8 residual blocks, meaning 16 convolutional layers overall. Further increasing the size did not affect performance significantly. One reason for that may be the size of the dataset. During the data-preprocessing-related changes, we came to the conclusion that taking into account the time dimension (basically the movement or dynamic patterns of the heart) had a major effect on the results (over six percent in the case of the short-axis views). Therefore, we created a 3D convolution-based ResNet model to properly handle the time dimension. We formed a 3D image, as time became the depth dimension of the image. This model performed better and more robustly (regarding the sensitivity for the hyper-parameters). However, the drawback of the 3D ResNet lies in its slow training speed. As the performance on the short-axis view was worse, we tried to increase the model size for this view only, but this did not cause relevant changes. Finally, we used the same architecture for all the views.

Data preprocessing. Data preprocessing and the input representation to the network proved to be the most important factors. To speed up the training, we tried less input data first. We used only two images from the long-axis views, one from the systole phase and one from the diastole phases. We used six images from the short-axis view and three images at the systole and the diastole phases, respectively. This input formation resulted in fair accuracy values (around 84 percent), but it turned out that taking images from other points of the cardiac cycle contributed to better results. Standardization (see Appendix A) had a very important role in achieving the final results. We identified the long-axis views to be noisy as a result of the different orientations of the images. This was not true for the short-axis. One way to cope with this is to use random rotation for augmentation with degrees between 0 and 180. We found this approach to be inefficient in helping the learning process. The standardization method caused a significant performance growth. Therefore, we used only a small eight-degree angle for rotation during augmentation. We also used cropping and some noise during augmentation.

Hyper-parameter tuning. When a model and a data preprocessing method were chosen, there were some hyper-parameters to optimize. These were batch size, number of epochs, learning rate, optimization algorithm, loss function, regularization method and their parameters, and the cropping size of the image. We chose batch size 16 because 8 was too noisy for the training. Larger batch sizes require too much memory. The number of epochs was chosen between 20 and 50, and we used early stopping to avoid overfitting. We found that the AdamW [40] algorithm with learning rate 5×10^{-4} achieved better results than Adam, SGD, and RMSProp. We used focal loss [41], because focal loss can distinguish the easy samples from the difficult ones by applying a factor $((1-p)^{\gamma})$, which reduces the loss for the well-classified samples. Our intuition was that the samples contained some very difficult cases (due to etiologies such as amyloidosis, which is difficult to diagnose), and therefore, focal loss could help. In our experiments, we experienced L1 and L2 losses to be harmful, and dropout with large values was disadvantageous. This can be explained by the observation that batch normalization has some regularization effect, which can eliminate the need for dropout [42], and our 3D ResNet contains batch normalization layers. The final cropping size of the input image proved to be 150×150. Smaller (120×120) and larger sizes (190×190) were worse. For the larger size, the image can contain too much noise, while the smaller crop can miss some details with the heart not always being at the center of the image.

4. Discussion and Conclusions

Cardiovascular diseases are the leading causes of death around the world [1,2,43]. LVH is a well-recognized independent risk factor for several cardiovascular complications [5]. The diagnosis of LVH can be challenging. For this, there are some methods used in clinical practice such as electrocardiography, echocardiography, and CMR. CMR is a non-invasive tool for diagnosing myocardial pathologies. CMR-based hypertrophy detection can be more efficient and reliable and may improve the diagnostic method in order to recognize LVH in an earlier stage. We developed a deep-learning-based algorithm for identifying left ventricular hypertrophy during a CMR examination (on-site) and for helping the diagnostic process following the examination (off-site). The on-site application can save time, if the algorithm indicates the presence of LVH right after the long-axis measurements; therefore, some additional, necessary images could be acquired and contrast administration should be applied. With the use of on-site application, the CMR protocol can be changed during the scanning, in order to avoid the need to call back the patient for an additional CMR examination to provide the correct diagnosis. Nevertheless, if the algorithm is used during post-process evaluation, it can warn the reader that LVH is present, so the diagnostic accuracy can be improved. This is important because the identification of the incipient or milder form of LVH is difficult for less-experienced readers, and early detection of LVH and subsequent therapy are key factors in reducing cardiovascular morbidity and mortality [38,44]. Our algorithm achieved a performance close to the medical experts' (readers) scores. Our comparison was based on the F1-score, precision, and recall. The model we implemented was an ensemble model. Each view had a separate extractor, and the features extracted from the acquired images were concatenated. Then, an ensemble classifier takes the concatenated features as the input and calculates the probability of having LVH. The dataset was collected from the Heart and Vascular Center of Semmelweis University, and it contains the raw image scans with all available views (long-axis and short-axis cine images) and the corresponding pathologies.

Our algorithm had a recall rate of 90% when the combination of long-axis views was used as the input. In the case of the combination of long-axis views and short-axis views, we had a 96% rate. The corresponding F1-scores were 89% and 91%, respectively. High recall is beneficial, because fewer LVH cases will be left undiagnosed. False positives (predicted as LVH, yet normal) can be discarded by the experts supervising the examination. In order to judge the applicability of our method, we established a baseline by measuring the scores of medical experts. The measurement involved six readers with varying levels of experience. The measurement was designed to simulate a realistic clinical scenario where the reader has no clear reference to the real case, but has access to the images of full CMR scans. To make it more realistic, we included several other diseases in addition to LVH. We included diseases that appear frequently in clinical practice, and the readers were blinded to the purpose of the study. There are three main outcomes of the human experiment: (1) the differences among the scores (F1-score, recall, etc.) of the readers were surprisingly small; (2) recall was the highest value indicating that the readers had a bias toward having a cardiac disease; (3) we obtained the baseline values for the scores (F1-score—95%, recall—98%); see Table 2. High recall was also achieved by our algorithm in the case of the combined long-axis and short-axis model. Figures 5 and 6 indicate that our algorithm can already be advantageous in clinical practice even though there is still room for improvement.

We claim that by using a larger dataset, the gap can be bridged and that this method can be a good candidate to become part of the daily clinical routine during CMR examinations. Our method was limited to only one vendor and clinic center. For creating a more robust method, the model should be trained on data gathered from different clinic centers and vendors. Another limitation is the classification of etiologies. The current method differentiates between two groups, normal (healthy) subjects and hypertrophy. There are different etiologies for hypertrophy (e.g., HCM, amyloidosis), which can be differentiated by including late enhancement images. From the dataset, we excluded healthy athletes, but LVH can be present as a physiological condition in athlete's heart; therefore, it could be an interesting topic to differentiate between physiologic and pathologic LVH.

To the best of our knowledge, this is the first paper where a method for automatic classification of LVH from different CMR images (short-axis, long-axis cine images) was investigated and compared to medical experts. Future work can focus on the separation of the etiologies within LVH automatically. Sports-related LVH should be also addressed in order to create a more complete methodology.

Author Contributions: Conceptualization, A.B., F.I.S. and H.V.; methodology, A.B. and F.I.S.; software, A.B.; validation, A.B. and F.I.S.; resources, K.C., H.V. and B.M.; data curation, A.B., F.I.S., Z.D. and L.S.; writing—original draft preparation, A.B. and F.I.S.; writing—review and editing, A.B., F.I.S., K.C. and H.V.; visualization, A.B. and F.I.S.; supervision, K.C., H.V. and B.M.; funding acquisition, K.C., H.V. and M.B. All authors have read and agreed to the published version of the manuscript.

Funding: This study was supported by the National Research, Development and Innovation Office of Hungary (NKFIA; 019-1.1.1-PIACI-KFI-2019-00263). This study was also supported by the National Research, Development and Innovation Office of Hungary (NKFIA; NVKP-16-1-2016-0017). The study was financed by the Research Excellence Programme of the Ministry for Innovation and Technology in Hungary within the framework of the Bioimaging Thematic Programme of Semmelweis University and by the Ministry of Innovation and Technology NRDI Office within the framework of the Artificial

Intelligence National Laboratory Program. Project no. TKP2021-NKTA-46 has been implemented with the support provided by the Ministry of Innovation and Technology of Hungary from the National Research, Development and Innovation Fund, financed under the TKP2021-NKTA funding scheme.

Institutional Review Board Statement: The study was conducted according to the guidelines of the Declaration of Helsinki and approved by the Institutional Review Board (or Ethics Committee) of The Heart and Vascular Center of Semmelweis University (Protocol Code NVKP, date: 5 June 2020).

Informed Consent Statement: Informed consent was obtained from all subjects involved in the study.

Data Availability Statement: The data presented in this study are available upon request from the corresponding author. The data are not publicly available; for acquiring the dataset, the permission of the local institution board is necessary.

Conflicts of Interest: The authors declare that they have no known competing financial interests or personal relationships that could have appeared to influence the work reported in this paper.

Appendix A. Standardization

The method for standardizing the long-axis images is based on defining a reference system relative to a fixed axis. This axis is the Z-axis, which points from the feet to the head of the patient, running parallel with the bore of the MRI. Each acquired image has a plane, and it can be described in this coordinate system. The plane is characterized by its normal vector and its orientation. The orientation is relative to the Z-axis. The images are stored in dicom files, which contain the orientation and position matrices. The standardization is achieved by rotation with a proper angle around the axis parallel with the normal vector and crossing the middle of the image; see Figure A1.

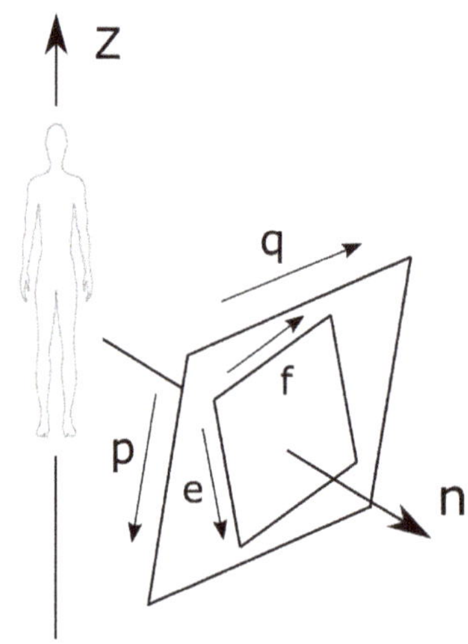

Figure A1. Schematic illustration of the vectors used in the explanation of the standardization process.

First, the algorithm calculates the normal vector of the image from the orientation matrix. The orientation matrix contains the directions of the left side and the upper side of the image ($\vec{e}$ and $\vec{f}$). Therefore, the normal vector is:

$$\vec{n} = \vec{e} \times \vec{f}. \tag{A1}$$

The normal vectors are almost the same for each view. Then, a new reference frame can be calculated ($\vec{p}$ and $\vec{q}$):

$$\vec{q} = \vec{z} \times \vec{n}, \ \vec{p} = \vec{q} \times \vec{n} \tag{A2}$$

where $\vec{z} = (0, 0, 1)$, then $\vec{p}$, $\vec{q}$ are normalized. The orientation is defined as the direction of $\vec{e}$ in the $\vec{p}$, $\vec{q}$ plane:

$$\vec{d} = [\vec{e} \cdot \vec{p}, \vec{e} \cdot \vec{q}]. \tag{A3}$$

We can define a reference orientation $(\vec{d_0})$, then each image can be compared and rotated against the reference orientation. To decrease the size of the required rotation angle, we calculated the average orientation of the images in the dataset per view. Then, we defined the reference orientations according to the average values. For the sake of completeness, these values were: LA2 (−0.937, 0.166), LA4 (0.632, 0.032) and LALVOT (−0.0054, −0.635). The rotation angle (φ) is given as follows:

$$\cos \varphi = \vec{d} \cdot \vec{d_0}. \tag{A4}$$

Appendix B. Parameters

Table A1. The hyper-parameters used in the best-preforming model.

Parameter	Value
batch size	16
learning rate	0.0005
optimizer	AdamW
input shape	150 × 150
max angle	8

Appendix C. Example Images

The following images show examples for different heart conditions: normal, HCM, amyloidosis, and Anderson–Fabry disease. In each row of pictures, the views from left to right are the following: short-axis, long-axis 2-chamber, long-axis 4-chamber, and long-axis 3 chambers-view.

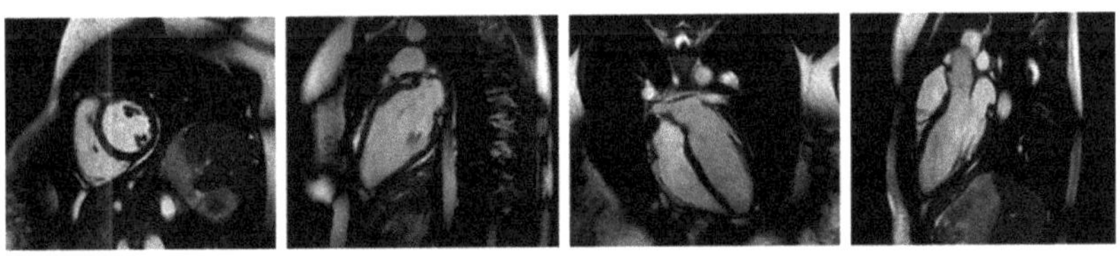

Figure A2. Short-axis cine image and long-axis cine images of healthy subject without left ventricular hypertrophy.

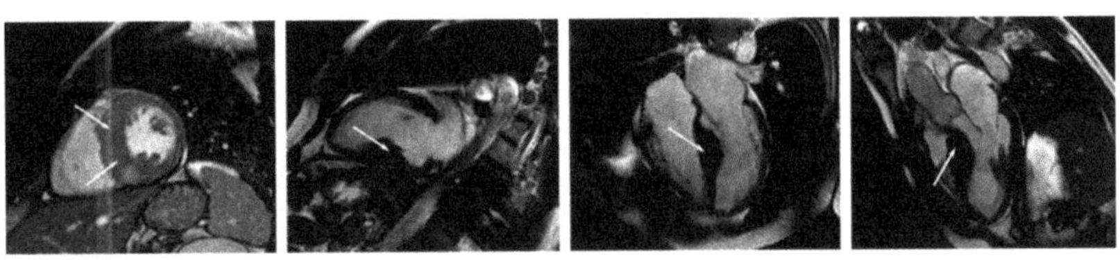

Figure A3. Short-axis cine image and long-axis cine images demonstrate left ventricular hypertrophy in a patient with hypertrophic cardiomyopathy (HCM). Cine images show marked asymmetrical septal hypertrophy (white arrows) corresponding with HCM.

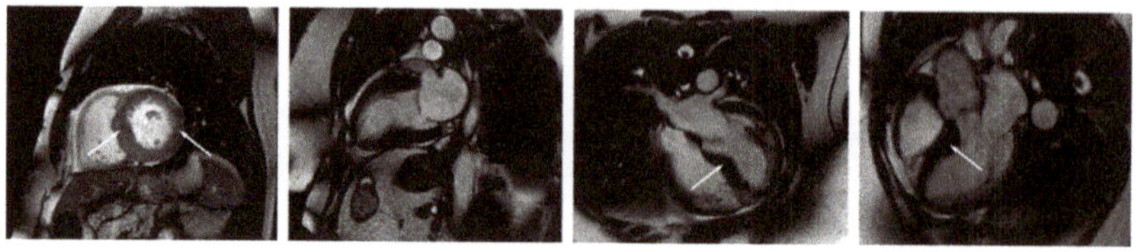

Figure A4. Short-axis cine image and long-axis cine images demonstrate concentric left ventricular hypertrophy with subtle septal predominance (white arrows), in a patient with Anderson–Fabry disease.

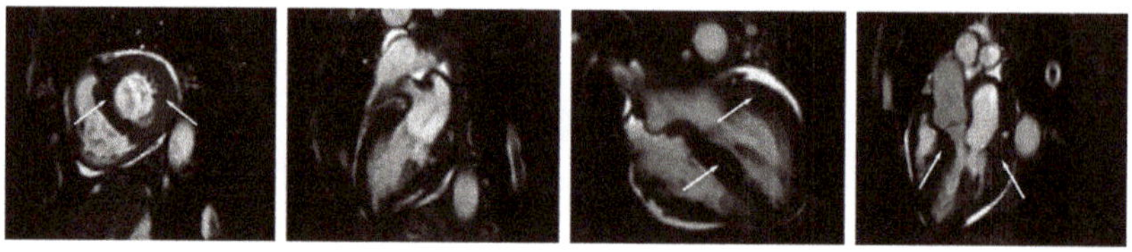

Figure A5. Short-axis cine image and long-axis cine images show marked, concentric left ventricular hypertrophy (white arrows) in a patient with cardiac amyloidosis.

References

1. Namara, K.; Alzubaidi, H.; Jackson, J.K. Cardiovascular disease as a leading cause of death: How are pharmacists getting involved? *Integr. Pharm. Res. Pr.* **2019**, *8*, 1–11. [CrossRef] [PubMed]
2. Roth, G.A.; Mensah, G.A.; Johnson, C.O.; Addolorato, G.; Ammirati, E.; Baddour, L.M.; Barengo, N.C.; Beaton, A.Z.; Benjamin, E.J.; Benziger, C.P.; et al. Global Burden of Cardiovascular Diseases and Risk Factors. *J. Am. Coll. Cardiol.* **2021**, *77*, 1958–1959. [CrossRef]
3. Foley, J.R.; Plein, S.; Greenwood, J.P. Assessment of stable coronary artery disease by cardiovascular magnetic resonance imaging: Current and emerging techniques. *World J. Cardiol.* **2017**, *9*, 92–108. [CrossRef] [PubMed]
4. Cuspidi, C.; Sala, C.; Negri, F.; Mancia, G.; Morganti, A. Prevalence of left ventricular hypertrophy in hypertension: An updated review of echocardiographic studies. *J. Hum. Hypertens.* **2012**, *26*, 343–349. [CrossRef]
5. Bluemke, D.A.; Kronmal, R.A.; Lima, J.A.; Liu, K.; Olson, J.; Burke, G.L.; Folsom, A.R. The relationship of left ventricular mass and geometry to incident cardiovascular events: The MESA (Multi-Ethnic Study of Atherosclerosis) study. *J. Am. Coll. Cardiol.* **2008**, *52*, 2148–2155. [CrossRef]
6. Kawel-Boehm, N.; Kronmal, R.; Eng, J.; Folsom, A.; Burke, G.; Carr, J.J.; Shea, S.; Lima, J.A.; Bluemke, D.A. Left Ventricular Mass at MRI and Long-term Risk of Cardiovascular Events: The Multi-Ethnic Study of Atherosclerosis (MESA). *Radiology* **2019**, *293*, 107–114. [CrossRef]
7. Devereux, R.B. Is the electrocardiogram still useful for detection of left ventricular hypertrophy? *Circulation* **1990**, *81*, 1144–6. [CrossRef]
8. Goldberger, A.; Goldberger, Z.; Shvilkin, A. *Goldberger's Clinical Electrocardiography: A Simplified Approach*, 9th ed.; Elsevier: Amsterdam, The Netherlands, 2017.
9. Jain, A.; Tandri, H.; Dalal, D.; Chahal, H.; Soliman, E.; Prineas, R.; Folsom, A.; Lima, J.; Bluemke, D. Diagnostic and prognostic utility of electrocardiography for left ventricular hypertrophy defined by magnetic resonance imaging in relationship to ethnicity: The Multi-Ethnic Study of Atherosclerosis (MESA). *Am. Heart J.* **2010**, *159*, 652–658. [CrossRef]
10. Okin, P.; Devereux, R.; Harris, K.; Jern, S.; Kjeldsen, S.; Julius, S.; Edelman, J.; Dahlöf, B.; Investigators, L.S. Regression of electrocardiographic left ventricular hypertrophy is associated with less hospitalization for heart failure in hypertensive patients. *Ann. Intern. Med.* **2007**, *147*, 311–319. [CrossRef]
11. Foppa, M.; Duncan, B.; Rohde, L. Echocardiography-based left ventricular mass estimation. How should we define hypertrophy? *Cardiovasc. Ultrasound* **2005**, *3*, 17. [CrossRef]
12. Grajewski, K.G.; Stojanovska, J.; Ibrahim, E.S.H.; Sayyouh, M.; Attili, A. Left Ventricular Hypertrophy: Evaluation With Cardiac MRI. *Curr. Probl. Diagn. Radiol.* **2020**, *49*, 460–475. [CrossRef] [PubMed]
13. Semsarian, C.; Ingles, J.; Maron, M.; Maron, B. New perspectives on the prevalence of hypertrophic cardiomyopathy. *J. Am. Coll. Cardiol.* **2015**, *65*, 1249–1254. [CrossRef] [PubMed]
14. Maron, B.J.; Maron, M.S. Hypertrophic cardiomyopathy. *Lancet* **2013**, *381*, 242–255. [CrossRef]
15. Maron, B.J.; Ommen, S.R.; Semsarian, C.; Spirito, P.; Olivotto, I.; Maron, M.S. Hypertrophic Cardiomyopathy: Present and Future, With Translation Into Contemporary Cardiovascular Medicine. *J. Am. Coll. Cardiol.* **2014**, *64*, 83–99. [CrossRef]
16. Bai, W.E.A. Automated cardiovascular magnetic resonance image analysis with fully convolutional networks. *J. Cardiovasc. Magn. Reson.* **2018**, *20*, 65. [CrossRef]

17. Bernard, O.; Lal, E.A.; Zotti, C.; Cervenansky, F.; Yang, X.; Heng, P.A.; Cetin, I.; Lekadir, K.; Camara, O.; Ballester, M.A.G.; et al. Deep Learning Techniques for Automatic MRI Cardiac Multi-structures Segmentation and Diagnosis: Is the Problem Solved? *IEEE Trans. Med. Imaging* **2018**, *37*, 2514–2525. [CrossRef] [PubMed]
18. Lu, Y.; Connelly, K.; Dick, A.; Wright, G.; Radau, P. Automatic functional analysis of left ventricle in cardiac cine MRI. *Quant. Imaging Med. Surg.* **2013**, *3*, 200–209. [CrossRef]
19. Tao, Q.; Yan, W.; Wang, Y.; Paiman, E.H.; Shamonin, D.P.; Garg, P.; Plein, S.; Huang, L.; Xia, L.; Sramko, M.; et al. Deep Learning–based Method for Fully Automatic Quantification of Left Ventricle Function from Cine MR Images: A Multivendor, Multicenter Study. *Radiology* **2019**, *290*, 81–88. [CrossRef]
20. Gopalakrishnan, V.; Menon, P.G.; Madan, S. cMRI-BED: A novel informatics framework for cardiac MRI biomarker extraction and discovery applied to pediatric cardiomyopathy classification. *BioMed. Eng. OnLine* **2015**, *14*, S7. [CrossRef]
21. Mantilla, J.; Garreau, M.; Bellanger, J.J.; Paredes, J.L. Machine Learning Techniques for LV Wall Motion Classification Based on Spatio-temporal Profiles from Cardiac Cine MRI. In Proceedings of the 2013 12th International Conference on Machine Learning and Applications, Miami, FL, USA, 4–7 December 2013; Volunme 1, pp. 167–172. [CrossRef]
22. Chen, R.; Lu, A.; Wang, J.; Ma, X.; Zhao, L.; Wu, W.; Du, Z.; Fei, H.; Lin, Q.; Yu, Z.; et al. Using machine learning to predict one-year cardiovascular events in patients with severe dilated cardiomyopathy. *Eur. J. Radiol.* **2019**, *117*, 178–183. [CrossRef]
23. Curiale, A.H.; Colavecchia, F.D.; Kaluza, P.; Isoardi, R.A.; Mato, G. Automatic myocardial segmentation by using a deep learning network in cardiac MRI. In Proceedings of the 2017 XLIII Latin American Computer Conference (CLEI), Córdoba, Argentina, 4–8 September 2017; pp. 1–6. [CrossRef]
24. Winther, H.B.; Hundt, C.; Schmidt, B.; Czerner, C.; Bauersachs, J.; Wacker, F.; Vogel-Claussen, J. ν-net: Deep Learning for Generalized Biventricular Cardiac Mass and Function Parameters. *arXiv* **2017**, arXiv:1706.04397.
25. Vigneault, D.M.; Xie, W.; Ho, C.Y.; Bluemke, D.A.; Noble, J.A. Omega-Net: Fully Automatic, Multi-View Cardiac MR Detection, Orientation, and Segmentation with Deep Neural Networks. *arXiv* **2017**, arXiv:1711.01094.
26. Radau, P.; Lu, Y.; Connelly, K.; Paul, G.; Dick, A.; Wright, G. Evaluation Framework for Algorithms Segmenting Short Axis Cardiac MRI. *MIDAS J.* **2009**, *49*. [CrossRef]
27. Suinesiaputra, A.; Cowan, B.R.; Al-Agamy, A.O.; Elattar, M.A.; Ayache, N.; Fahmy, A.S.; Khalifa, A.M.; Medrano-Gracia, P.; Jolly, M.P.; Kadish, A.H.; et al. A collaborative resource to build consensus for automated left ventricular segmentation of cardiac MR images. *Med. Image Anal.* **2014**, *18*, 50–62. [CrossRef] [PubMed]
28. Martin-Isla, C.; Campello, V.M.; Izquierdo, C.; Raisi-Estabragh, Z.; Baeßler, B.; Petersen, S.E.; Lekadir, K. Image-Based Cardiac Diagnosis With Machine Learning: A Review. *Front. Cardiovasc. Med.* **2020**, *7*, 1. [CrossRef] [PubMed]
29. Madani, A.; Ong, J.R.; Tibrewal, A.; Mofrad, M.R.K. Deep echocardiography: Data-efficient supervised and semi-supervised deep learning towards automated diagnosis of cardiac disease. *Npj. Digit. Med.* **2018**, *1*, 59. [CrossRef] [PubMed]
30. Kwon, J.M.; Jeon, K.H.; Kim, H.M.; Kim, M.J.; Lim, S.M.; Kim, K.H.; Song, P.S.; Park, J.; Choi, R.K.; Oh, B.H. Comparing the performance of artificial intelligence and conventional diagnosis criteria for detecting left ventricular hypertrophy using electrocardiography. *EP Eur.* **2020**, *22*, 412–419. [CrossRef]
31. Jothiramalingam, R.; Jude, A.; Patan, R.; Ramachandran, M.; Duraisamy, J.H.; Gandomi, A.H. Machine learning-based left ventricular hypertrophy detection using multi-lead ECG signal. *Neural Comput. Appl.* **2021**, *33*, 4445–4455. [CrossRef]
32. Jian, Z.; Wang, X.; Zhang, J.; Wang, X.; Deng, Y. Diagnosis of left ventricular hypertrophy using convolutional neural network. *BMC Med. Inform. Decis. Mak.* **2020**, *20*, 243. [CrossRef]
33. De la Garza-Salazar, F.; Romero-Ibarguengoitia, M.E.; Rodriguez-Diaz, E.A.; Azpiri-Lopez, J.R.; González-Cantu, A. Improvement of electrocardiographic diagnostic accuracy of left ventricular hypertrophy using a Machine Learning approach. *PLoS ONE* **2020**, *15*, e0232657. [CrossRef]
34. Massera, D.; McClelland, R.L.; Ambale-Venkatesh, B.; Gomes, A.S.; Hundley, W.G.; Kawel-Boehm, N.; Yoneyama, K.; Owens, D.S.; Garcia, M.J.; Sherrid, M.V.; et al. Prevalence of Unexplained Left Ventricular Hypertrophy by Cardiac Magnetic Resonance Imaging in MESA. *J. Am. Heart Assoc.* **2019**, *8*, e012250. [CrossRef] [PubMed]
35. Khened, M.; Alex, V.; Krishnamurthi, G. Densely Connected Fully Convolutional Network for Short-Axis Cardiac Cine MR Image Segmentation and Heart Diagnosis Using Random Forest. Statistical Atlases and Computational Models of the Heart. In *ACDC and MMWHS Challenges*; Pop, M., Sermesant, M., Jodoin, P.M., Lalande, A., Zhuang, X., Yang, G., Young, A., Bernard, O., Eds.; Springer International Publishing: Cham, Switzerland, 2018; pp. 140–151.
36. Augusto, J.B.; Davies, R.H.; Bhuva, A.N.; Knott, K.D.; Seraphim, A.; Alfarih, M.; Lau, C.; Hughes, R.K.; Lopes, L.R.; Shiwani, H.; et al. Diagnosis and risk stratification in hypertrophic cardiomyopathy using machine learning wall thickness measurement: A comparison with human test-retest performance. *Lancet Digit. Health* **2021**, *3*, e20–e28. [CrossRef]
37. Tobon-Gomez, C.; Butakoff, C.; Yushkevich, P.; Huguet, M.; Frangi, A.F. 3D Mesh Based Wall Thickness Measurement: Identification of Left Ventricular Hypertrophy Phenotypes. In Proceedings of the 2nd Annual International Conference of the IEEE EMBS, Buenos Aires, Argentina, 31 August–4 September 2010. [CrossRef]
38. Milan, A.; Caserta, M.A.; Avenatti, E.; Abram, S.; Veglio, F. Anti-hypertensive drugs and left ventricular hypertrophy: A clinical update. *Intern. Emerg. Med.* **2010**, *5*, 469–479. [CrossRef] [PubMed]
39. Budai, A.; Suhai, F.I.; Csorba, K.; Toth, A.; Szabo, L.; Vago, H.; Merkely, B. Fully automatic segmentation of right and left ventricle on short-axis cardiac MRI images. *Comput. Med. Imaging Graph.* **2020**, *85*, 101786. [CrossRef]
40. Loshchilov, I.; Hutter, F. Decoupled Weight Decay Regularization. *arXiv* **2019**, arXiv:1711.05101.

41. Lin, T.; Goyal, P.; Girshick, R.B.; He, K.; Dollár, P. Focal Loss for Dense Object Detection. *arXiv* **2017**, arXiv:1708.02002.
42. Ioffe, S.; Szegedy, C. Batch Normalization: Accelerating Deep Network Training by Reducing Internal Covariate Shift. *arXiv* **2015**, arXiv:1502.03167.
43. Islam, M.M.; Beverung, S.; Steward, R., Jr. Bio-Inspired Microdevices that Mimic the Human Vasculature. *Micromachines* **2017**, *8*, 299. [CrossRef]
44. de Hartog-Keyzer, J.M.L.; El Messaoudi, S.; Harskamp, R.; Vart, P.; Ringoir, L.; Pop, V.; Nijveldt, R. Electrocardiography for the detection of left ventricular hypertrophy in an elderly population with long-standing hypertension in primary care: A secondary analysis of the CHELLO cohort study. *Cardiovasc. Ultrasound* **2020**, *10*, e038824. [CrossRef]

applied sciences

MDPI

Article

Pulmonary Artery Remodeling and Advanced Hemodynamics: Magnetic Resonance Imaging Biomarkers of Pulmonary Hypertension

Zachary M. Hong [1,2,3,4] and Julio Garcia [1,2,3,4,5,*]

1 Department of Cardiac Sciences, University of Calgary, Calgary, AB T2N 1N4, Canada; zachary.hong@ucalgary.ca
2 Department of Radiology, University of Calgary, Calgary, AB T2N 1N4, Canada
3 Stephenson Cardiac Imaging Centre, University of Calgary, Calgary, AB T2N 1N4, Canada
4 Libin Cardiovascular Institute, University of Calgary, Calgary, AB T2N 1N4, Canada
5 Alberta Children's Hospital Research Institute, University of Calgary, Calgary, AB T2N 1N4, Canada
* Correspondence: julio.garciaflores@ucalgary.ca

Featured Application: Pulmonary artery remodeling and 3D flow biomarkers can be useful to characterize pulmonary hypertension severity and progression.

Abstract: Poorly characterized by non-invasive diagnostic imaging techniques, pulmonary hypertension (PHT) is commonly associated with changes in vascular hemodynamics and remodeling of pulmonary artery architecture. These disease phenotypes represent potential biomarkers of interest in clinical environment. In this retrospective clinical study, 33 patients with pulmonary hypertension and seventeen controls were recruited. Architectural remodeling was characterized using 3D-contrast enhanced angiogram via the measurement of pulmonary artery diameters, bifurcation distances, and angles. Hemodynamics were characterized using 4D-flow magnetic resonance imaging (MRI) via wall shear stress, kinetic energy, vorticity, and directional flow dynamics. Parameters were compared using independent samples student's t-tests. Correlational analysis was performed using Pearson's correlation. PHT patients demonstrated dilation in the main and right branch of the pulmonary artery ($p < 0.05$). Furthermore, these patients also exhibited increases in bifurcation distances in the left and right pulmonary arteries ($p < 0.05$). Wall shear stress, maximum kinetic energy, and energy loss were decreased in the pulmonary artery ($p < 0.001$). Correlations were observed between peak velocities and right ventricle ejection fraction ($r = 0.527$, $p < 0.05$). These findings suggest that pulmonary artery remodeling and hemodynamic changes may possess clinical utility as MRI biomarkers for PHT.

Keywords: pulmonary hypertension; magnetic resonance imaging; 4D-flow MRI; heart hemodynamics

Citation: Hong, Z.M.; Garcia, J. Pulmonary Artery Remodeling and Advanced Hemodynamics: Magnetic Resonance Imaging Biomarkers of Pulmonary Hypertension. *Appl. Sci.* **2022**, *12*, 3518. https://doi.org/10.3390/app12073518

Academic Editor: Vladislav Toronov

Received: 23 February 2022
Accepted: 29 March 2022
Published: 30 March 2022

1. Introduction

Pulmonary hypertension (PHT) is a complex pathophysiological disorder characterized by an elevation in pulmonary arterial pressure (PAP). This diagnosis is given when mean PAP exceeds 25 mmHg at rest using invasive right heart catheterization [1]. In early stages of the disease, elevations in PAP and architectural remodeling typically induce dilation and overload-induced hypertrophy of the right ventricle. Over time, this results in impairment of cardiac function, eventual right heart failure, and secondary pathophysiology such as right atrial dilation [2]. Although two-dimensional (2D) Doppler echocardiography is commonly used to provide evidence of the disease hallmarks in the right ventricle, and right ventricular catherization is needed for a conclusive diagnosis. Thus, other non-invasive diagnostic imaging biomarkers would be beneficial in clinical settings [3]. Magnetic resonance imaging (MRI) and angiography (MRA) could address this unmet need. Relative to 2D Doppler echocardiography, MRI possesses two advantages.

First, it can perform multiplanar analysis allowing for accurate volumetric measurement of the vasculature. Second, it can evaluate hemodynamic changes in the pulmonary pathway via 4D-flow MRI. In both cases, volume and flow measurements demonstrate low variability and high test–re-test reliability [3,4]. Although MRI has improved the characterization of right ventricular pathophysiology, it has also identified potential biomarkers of the disease in the pulmonary artery (PA). Patients diagnosed with PHT often exhibit hemodynamic changes such as decreased vorticity, peak velocity, and wall shear stress (WSS) [5–7]. Furthermore, anatomic remodeling, including PA dilation and medial thickening, can be simultaneously observed [8]. Although promising, these novel biomarkers and their relationship must be investigated.

The purpose of this study was to test the hypothesis that hemodynamic changes within the pulmonary artery are associated with remodeling of the surrounding vascular architecture. In addition, we aimed to further characterize PHT biomarkers in the PA by investigating novel 3D flow biomarkers and bifurcation distances.

2. Materials and Methods

2.1. Study Population

A total of 51 subjects were recruited retrospectively, including 17 PHT patients and 33 controls. All subjects were registered within the Cardiovascular Imaging Registry of Calgary (CIROC). The study was approved by the University of Calgary Research Ethics Board and all subjects provided written informed consent. All research activities were performed in accordance with the Declaration of Helsinki. The study was coordinated by commercial software (cardioDI^{TM}, Cohesic Inc., Calgary, AB, Canada) for the routine capture of patient informed consent, for health questionnaires, and for standardized collection of MRI-related variables. PHT was clinically defined as a mean PA pressure $\geq$25 mmHg previously assessed by clinical invasive catheterization or echocardiography. Patients were recruited at the time of visit, with exclusions occurring for patients with prior surgical interventions in the pulmonary valve and/or artery; complex congenital heart disease; and/or contraindications to MRI. Control subjects ($\geq$18 years of age) underwent similar workflow and confirmed no prior history of cardiovascular disease with a certified nurse. No catherization was performed in healthy controls. Prior to scanning, demographic measurements including age, sex, height, weight, and heart rate were obtained. Volume and mass measurements were normalized to body surface area using the Mosteller formula.

2.2. Cardiac Magnetic Resonance Imaging Protocol

Cardiac imaging examination was performed using 3T MRI scanners (Skyra and Prima, Siemens, Erlangen, Germany). Indication-based protocolling ensured consistent imaging procedures for all subjects, and cardiac imaging was performed in accordance with published recommendations [9]. Standard routine retrospective electrocardiographic gating, and time-resolved balanced steady-state free precession (SSFP) cine imaging in four-chamber, three-chamber, two-chamber, and short-axis views of LV at end-expiration, was performed. Contrast usage of gadolinium contrast volume of 0.2 mmol/kg (Gadovist®, Bayer Inc., Mississauga, ON, Canada) was administrated to acquire a contrast-enhanced 3D magnetic resonance angiogram (CE MRA) of the cardiovascular structure. Time-resolved three-dimensional phase-contrast MRI with three-directional velocity encoding and retrospective ECG-gating (4D-flow, Siemens WIP 785A) was performed for 5–10 min, following contrast administration to measure in-vivo blood flow velocities within the whole heart. We have previously reported this whole-heart protocol [10–12]. Briefly, 4D-flow data was acquired during free breathing using navigator gating of diaphragmatic motion; sequence parameters were as follows: flip angle = 15 degrees, spatial resolution = 2.0–3.5 $\times$ 2.0–3.5 $\times$ 2.5–3.5 mm; temporal resolution = 39–48 ms; and velocity sensitivity = 150–250 cm/s. Total acquisition time varies between 5–10 min, depending on heart rate and respiratory navigator efficiency. The number of phases was adjusted to 25.

2.3. Standard Cardiac Imaging Analysis

Standard cardiac images were analyzed by a blinded reader to the study the same day of the acquisition using dedicated software cvi[42] version 5.11.5 (Circle Cardiovascular Imaging Inc, Calgary, AB, Canada) to determine left and right end-diastolic volume (LVEDV; RVEDV), LV and RV end-systolic volume (LVSEV; RVESV), and LV and RV ejection fraction (LVEF; RVEF). Furthermore, the contrast-enhanced angiogram (CE MRA) was used to measure pulmonary artery diameters (LPA, MPA, and RPA), bifurcation distances, and bifurcation angles (Figure 1D).

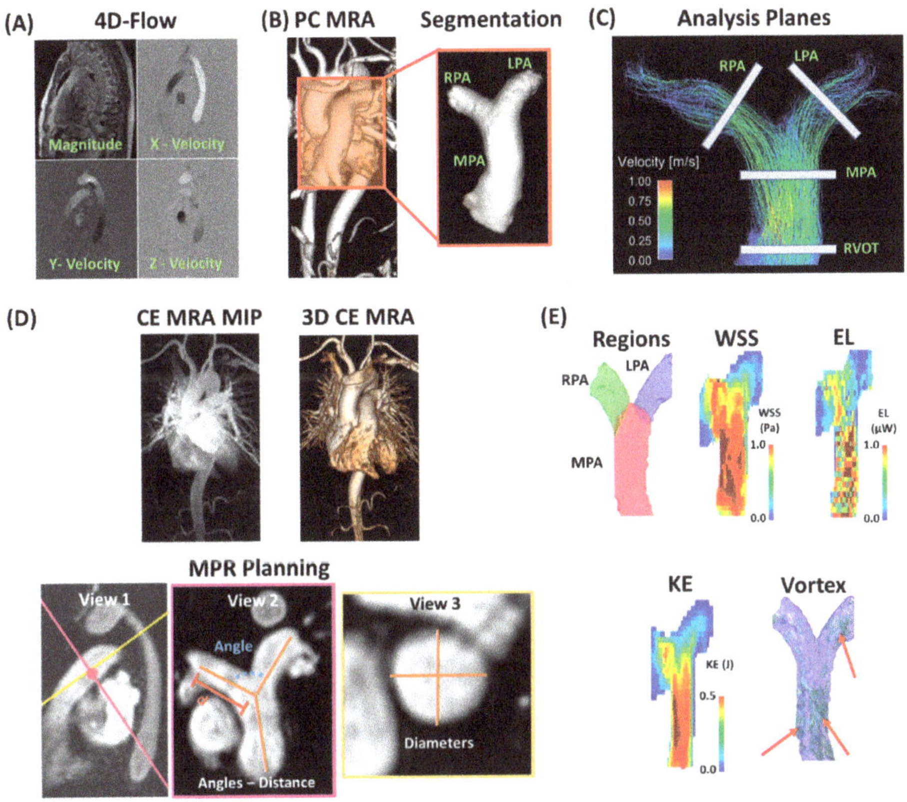

Figure 1. Data processing and analysis workflow. Panel (**A**) shows an example of corrected 4D-flow images. Panel (**B**) illustrates a time-average phase-contrast (PC) magnetic resonance angiogram (MRA) with an optimal threshold for visualizing the anatomy. The pulmonary artery (PA) was segmented by isolating the corresponding volume and identifying the main PA (MPA), right PA (RPA), and left PA (LPA). Panel (**C**) shows the 2D analysis planes for basic flow quantification in the right ventricle outflow track (RVOT), MPA, RPA, and LPA. Panel (**D**) shows contrast-enhanced MRA (CE MRA) maximum intensity projection (MIP) and the corresponding 3D volume of a sample case. Multi-planar reconstruction (MPR) was used for planning of angle (in blue), distance (d in red), and diameter (location in pink, diameters in golden). Pink and golden lines in View 1 represent the projection planes corresponding to View 2 and 3, respectively. Distance samples (orange lines in View 2) were defined for MPA the closer point to the pulmonary valve to the PA bifurcation; right and left PA distances were defined from the PA bifurcation to the closer visible in-plane ramification. Panel (**E**) shows examples of sub-volume regions of MPA, RPA, and LPA; wall shear stress (WSS); energy loss (EL); kinetic energy (KE); and vortex core size (red arrows point to vortex cores identified by λ_2).

2.4. 4D-Flow Data Analysis

All 4D-flow MRI data was pre-processed using in-house program developed in MATLAB 2020b (Mathworks, Natick, MA, USA), and the following tasks were performed: corrections for Maxwell terms, eddy currents, and aliasing, Figure 1A. After pre-processing, a 3D phase-contrast (PC) angiogram (PC MRA) was generated, Figure 1B. This angiogram was used to segment selected anatomical regions of the pulmonary artery, Figure 1B red inset, using in-house MATLAB-based tool "4D-Flow Analysis Tool" [13,14]. Analysis planes were created at the right ventricular outflow tract (RVOT), main pulmonary artery (MPA), the left pulmonary artery (LPA), and right pulmonary artery (RPA) using a specialized visualization software (Ensight 10.2, CEI Inc, Research Triangle Park, North Carolina, USA) to measure peak velocity, net flow, retrograde flow, regurgitation fraction, and forward flow, Figure 1C. Furthermore, the PC MRA was used to measure pulmonary artery diameters (LPA, MPA and RPA), bifurcation distances, and bifurcation angles (Figure 1D) in cvi[42] when standard CE MRA was not available in controls (n = 2). Volume sub-regions were created for local advanced 4D-flow analysis. These volume subregions were used to obtain the maximum velocities from the LPA, PA, and RPA using the approach proposed by Rose et al. [15].

Regional WSS was calculated at peak systole as previously described [16,17]. Energy loss (EL) was calculated as proposed by Barker et al. [18], which provides the rate of EL in a volume of interest (i.e., power in Watts) at a given time of the cardiac cycle, and peak systole was used in this study. Kinetic energy (KE) was calculated as reported by Geeraert et al. [13] in each sub-region at peak systole. Vortex size analysis at peak systole was performed as introduced by Garcia et al. [10] using the Lambda2 (λ_2) method, which is an accepted method to identify 3D vortices. Valid vortex networks were defined as those >75 mm^3 (considering a voxel resolution of 2.5 × 2.5 × 3.0 mm) with a shared connectivity neighborhood ≥4 voxels. This rule reduces the impact of possible incorrect derived vortices due to noise and/or partial derivative errors, which are directly impacted by 4D-flow spatial resolution. Examples for these advanced 4D-flow parameters are presented in Figure 1E.

2.5. Statistical Analysis

Statistical analysis was performed using SPSS 25 (SPSS, Chicago, IL, USA). Normality was assessed using normal plots and the Shapiro–Wilk test. Between-group comparisons for demographic, baseline cardiac MRI, hemodynamics, and anatomical measurements were performed using an independent samples t-test incorporating Levene's test for equality of variances. To assess the relationship between variables, a series of linear regression tests was performed using Pearson R-Values. If a significant relationship was found, a univariate linear model was generated using variables demonstrating statistical significance during the linear regression tests. Pearson correlation coefficients were also generated in relation to LVEF and RVEF. The significance threshold was <0.05.

3. Results

3.1. Patient Characteristics

Patient and control group characteristics inclusive of demographic and cardiac function data are summarized in Table 1. There were differences in age (55 ± 17 years vs. 39 ± 15 years, p = 0.001), sex (76% vs. 31% female, p = 0.002), height (1.64 ± 0.14 m vs. 1.73 ± 0.10 m, p = 0.023), and heart rate (76 ± 16 bpm vs. 66 ± 11 bpm, p = 0.016) between PHT patients and controls. BSA (1.88 ± 0.28 m^2 vs. 1.92 ± 0.34 m^2, p = 0.643) and weight (78 ± 18 kg vs. 79 ± 21 kg, p = 0.791) did not differ. Of the cardiac function parameters, PHT patients demonstrated lower LVEDV (127 ± 48 mL vs. 160 ± 35 mL, p = 0.011). However, other measurements, including LVEF (59 ± 13% vs. 61 ± 6%, p = 0.752), LVESV (58 ± 42 mL vs. 66 ± 23 mL, p = 0.381), RVEF (56 ± 7% vs. 55 ± 30%, p = 0.878), RVEDV (190 ± 79 mL vs. 173 ± 48 mL, p = 0.456), and RVESV (106 ± 72 mL vs. 77 ± 28 mL, p = 0.152), did not differ.

Table 1. Demographic, cardiac function baseline, and medications.

	Pulmonary Hypertension (n = 17)	Controls (n = 33)	p-Value
Age (years)	55 ± 17	39 ± 15	0.001
Sex (% female)	76	31	0.002
Height (m)	1.64 ± 0.14	1.73 ± 0.10	0.023
BSA (m^2)	1.88 ± 0.28	1.92 ± 0.34	0.643
Weight (kg)	78 ± 18	79 ± 21	0.791
Heart Rate (bpm)	76 ± 16	66 ± 11	0.016
Systolic Blood Pressure (mmHg)	114 ± 15	110 ± 16	0.396
Diastolic Blood Pressure (mmHg)	65 ± 10	63 ± 15	0.638
LVEF (%)	59 ± 13	61 ± 6	0.752
LVEDV (ml)	127 ± 48	160 ± 35	0.011
LVESV (ml)	58 ± 42	66 ± 23	0.381
RVEF (%)	55 ± 30	56 ± 7	0.878
RVEDV (ml)	190 ± 79	173 ± 48	0.456
RVESV (ml)	106 ± 72	77 ± 28	0.152
β–Blocker, n (%)	3 (18)		
Calcium channel blockers, n (%)	1 (6)		
Anticoagulants, n (%)	5 (29)		
Diuretics, n (%)	6 (35)		
ACE inhibitor, n (%)	4 (24)		
Statin, n (%)	2 (12)		
Digoxin, n (%)	1 (6)		

Data are expressed as mean ± SD. SD: standard deviation. LVEF: left ventricle ejection fraction. LVEDV: LV end-diastolic volume. LVESV: LV end-systolic volume. RVEF: right ventricle ejection fraction. RVEDV: RV end-diastolic volume. RVESV: RV end-systolic volume.

3.2. Anatomical Remodeling

Anatomical measurements of the PA are detailed in Table 2. Of the regions measured, only the MPA (AP: 32 ± 6 mm vs. 30 ± 4 mm, p = 0.050; RL: 32 ± 6 mm vs. 28 ± 3 mm, p = 0.005) and the RPA-AP (21 ± 4 mm vs. 19 ± 3 mm, p = 0.050) demonstrated dilatation in PHT patients. The LPA (AP: 20 ± 4 mm vs. 19 ± 4 mm, p = 0.370; CC: 21 ± 4 mm vs. 19 ± 3 mm, p = 0.226) and the RPA-CC (22 ± 4 mm vs. 21 ± 3 mm, p = 0.558) did not differ. The bifurcation distances of PHT patients were greater in the branches (LPA: 34 ± 6 mm vs. 29 ± 8 mm, p = 0.043; RPA: 34 ± 5 mm vs. 27 ± 6 mm, p < 0.001), but not the MPA (33 ± 5 vs. 30 ± 8 mm, p = 0.302) itself. Bifurcation angles did not differ (93 ± 9 degrees vs. 98 ± 15 degrees, p = 0.208).

Table 2. Anatomical measurements.

	Pulmonary Hypertension (n = 17)	Controls (n = 33)	p-Value
LPA Diameter			
AP (mm)	20 ± 4	19 ± 4	0.37
CC (mm)	21 ± 4	19 ± 3	0.226
MPA Diameter			
AP (mm)	32 ± 6	30 ± 4	0.05
RL (mm)	32 ± 6	28 ± 3	0.005
RPA Diameter			
AP (mm)	21 ± 4	19 ± 3	0.05
CC (mm)	22 ± 4	21 ± 3	0.558
Bifurcation Distance			
MPA (mm)	33 ± 5	30 ± 8	0.302
LPA (mm)	34 ± 6	29 ± 8	0.043
RPA (mm)	34 ± 5	27 ± 6	<0.001
Bifurcation Angle (deg)	93 ± 9	98 ± 15	0.208

Data are expressed as mean ± SD. SD: standard deviation. LPA: left pulmonary artery. MPA: main PA. RPA: right PA. deg: degrees. AP: anterior–posterior. CC: cranio-caudal.

3.3. Hemodynamic Characteristics

4D flow data is reported in Table 3, and sample datasets are illustrated in Figure 2. Maximum EL was lower in PHT patients in all regions of the PA (LPA: 1.06 ± 0.75 µW vs. 2.12 ± 1.22 µW, $p = 0.002$; MPA: 2.76 ± 1.20 µW vs. 4.30 ± 1.93 µW, $p = 0.001$; PA: 3.94 ± 1.71 µW vs. 7.27 ± 3.24 µW, $p < 0.001$; RPA: 0.88 ± 0.70 µW vs. 2.06 ± 1.22 µW, $p = 0.001$) relative to controls. Although mean EL did not differ in the MPA (1.24 ± 0.56 µW vs. 1.27 ± 0.45 µW, $p = 0.800$) or the entirety of the PA (1.76 ± 0.83 µW vs. 2.00 ± 1.09 µW, $p = 0.439$), it was increased in PHT patients in the LPA (1.76 ± 0.39 µW vs. 0.49 ± 0.51 µW, $p = 0.022$) and decreased in the RPA (0.06 ± 0.24 µW vs. 0.46 ± 0.56 µW, $p = 0.001$).

Table 3. 4D-Flow hemodynamic measurements.

	LPA			MPA			PA			RPA		
	Pulmonary Hypertension (n = 17)	Controls (n = 33)	p-Value	Pulmonary Hypertension (n = 17)	Controls (n = 33)	p-Value	Pulmonary Hypertension (n = 17)	Controls (n = 33)	p-Value	Pulmonary Hypertension (n = 17)	Controls (n = 33)	p-Value
Max EL (µW)	1.06 ± 0.75	2.12 ± 1.22	0.002	2.76 ± 1.20	4.30 ± 1.93	0.001	3.94 ± 1.71	7.27 ± 3.24	<0.001	0.88 ± 0.70	2.06 ± 1.22	0.001
Mean EL (µW)	1.76 ± 0.39	0.49 ± 0.51	0.022	1.24 ± 0.56	1.27 ± 0.45	0.8	1.76 ± 0.83	2.00 ± 1.09	0.439	0.06 ± 0.24	0.46 ± 0.56	0.001
Max WSS (Pa)	0.75 ± 0.29	1.17 ± 0.26	<0.001	0.97 ± 0.28	1.06 ± 0.18	0.224	0.98 ± 0.28	1.28 ± 0.28	0.001	0.64 ± 0.26	1.08 ± 0.30	<0.001
Mean WSS (Pa)	0.14 ± 0.05	0.19 ± 0.05	0.001	0.14 ± 0.03	0.17 ± 0.04	0.022	0.14 ± 0.03	0.18 ± 0.04	0.002	0.12 ± 0.04	0.17 ± 0.05	0.001
Mean WSS above 95% (Pa)	0.73 ± 0.28	1.14 ± 0.26	<0.001	0.94 ± 0.27	1.03 ± 0.17	0.234	0.96 ± 0.27	1.51 ± 1.44	0.123	0.63 ± 0.25	1.03 ± 0.32	<0.001
Mean Highest 5% WSS (Pa)	0.48 ± 0.20	0.76 ± 0.15	<0.001	0.52 ± 0.13	0.68 ± 0.14	<0.001	0.52 ± 0.14	0.74 ± 0.14	<0.001	0.40 ± 0.16	0.70 ± 0.18	<0.001
Max Vortex Volume (mm^3)	5.60 ± 3.01	4.44 ± 1.85	0.09	27.08 ± 10.66	20.33 ± 7.09	0.01	33.99 ± 12.14	27.47 ± 7.08	0.053	4.11 ± 2.22	3.70 ± 1.89	0.493
Mean Vortex Volume (mm^3)	3.55 ± 2.04	2.14 ± 1.11	0.014	19.00 ± 8.71	11.97 ± 4.76	0.005	24.55 ± 10.60	15.58 ± 5.41	0.004	2.61 ± 1.41	2.09 ± 1.19	0.175
Max KE (mJ)	1.1 ± 0.7	2.2 ± 1.2	<0.001	2.8 ± 1.1	4.2 ± 1.9	0.002	4.0 ± 1.7	7.3 ± 3.3	<0.001	0.7 ± 0.5	2.0 ± 1.2	<0.001

Data are expressed as mean ± SD. PA: pulmonary artery. LPA: left pulmonary artery. MPA: main PA. RPA: right PA. EL: energy loss. WSS: wall shear stress. KE: kinetic energy.

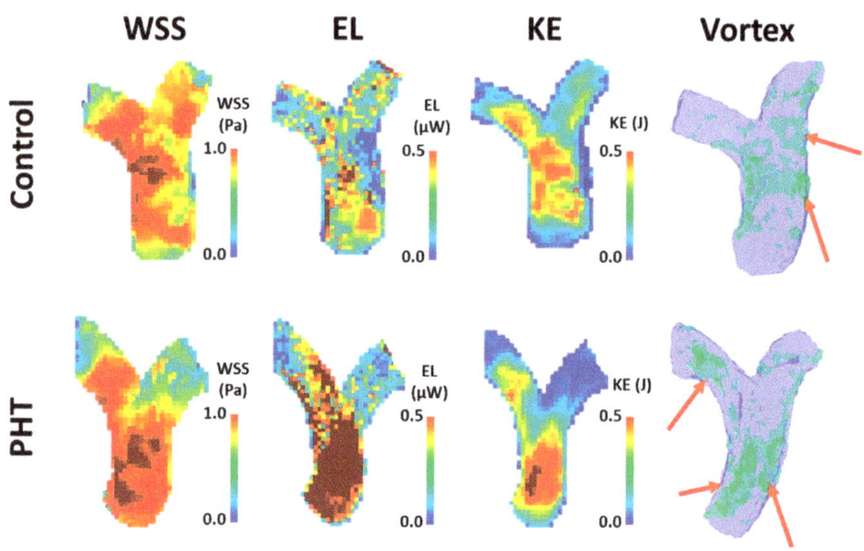

Figure 2. Sample dataset of advanced 4D-flow based measurements. Top raw shows a healthy control, man, 36-year-old. Bottom raw shows a pulmonary hypertension (PHT) patient, man, 58-year-old. WSS: wall shear stress; EL: energy loss; KE: kinetic energy.

Maximum WSS for PHT patients was lower in all regions (LPA: 0.75 ± 0.29 Pa vs. 1.17 ± 0.26 Pa, $p < 0.001$; PA: 0.98 ± 0.28 Pa vs. 1.28 ± 0.28 Pa, $p = 0.001$; RPA: 0.64 ± 0.26 Pa vs. 1.08 ± 0.30 Pa, $p < 0.001$), except the MPA (0.97 ± 0.28 Pa vs. 1.06 ± 0.18 Pa, $p = 0.224$). Mean WSS (LPA: 0.14 ± 0.05 Pa vs. 0.19 ± 0.05 Pa, $p = 0.001$; MPA: 0.14 ± 0.03 Pa vs. 0.18 ± 0.04 Pa, $p = 0.022$; PA: 0.14 ± 0.03 Pa vs. 0.18 ± 0.04 Pa, $p = 0.002$; RPA: 0.12 ± 0.04 Pa vs. 0.17 ± 0.05 Pa, $p < 0.001$) and the mean of the highest 5% WSS values (LPA: 0.48 ± 0.20 Pa vs. 0.76 ± 0.15 Pa, $p < 0.001$; MPA: 0.52 ± 0.13 Pa vs. 0.68 ± 0.14 Pa, $p < 0.001$; PA: 0.52 ± 0.14 Pa vs. 0.74 ± 0.14 Pa, $p < 0.001$; RPA: 0.40 ± 0.16 Pa vs. 0.70 ± 0.18 Pa, $p < 0.001$) were lower in all regions of the PA. The mean of all WSS values above 95% was decreased in the LPA (0.73 ± 0.28 Pa vs. 1.14 ± 0.26 Pa, $p < 0.001$) and RPA (0.63 ± 0.25 Pa vs. 1.03 ± 0.32 Pa, $p < 0.001$), but not the MPA (0.94 ± 0.27 Pa vs. 1.03 ± 0.17 Pa, $p = 0.234$) or the entirety of the PA (0.96 ± 0.27 Pa vs. 1.51 ± 1.44 Pa, $p = 0.123$) relative to controls. Maximum vortex volume was only increased in the MPA (27.08 ± 10.66 mm^3 vs. 20.33 ± 7.09 mm^3, $p = 0.01$) but otherwise did not differ (LPA: 5.60 ± 3.01 mm^3 vs. 4.44 ± 1.85 mm^3, $p = 0.090$; PA: 33.99 ± 12.14 mm^3 vs. 27.47 ± 7.08 mm^3, $p = 0.053$; RPA: 4.11 ± 2.22 mm^3 vs. 3.70 ± 1.89 mm^3, $p = 0.493$). Mean vortex volume was increased in all regions (LPA: 3.55 ± 2.04 mm^3 vs. 2.14 ± 1.11 mm^3, $p = 0.014$; MPA: 19.00 ± 8.71 mm^3 vs. 11.97 ± 4.76 mm^3, $p = 0.005$; PA: 24.55 ± 10.60 mm^3 vs. 15.58 ± 5.41 mm^3, $p = 0.004$) except for the RPA (2.61 ± 1.41 mm^3 vs. 2.09 ± 1.19 mm^3, $p = 0.175$). Maximum KE attained was also lower in PHT patients (LPA: 1.1 ± 0.7 mJ vs. 2.2 ± 1.2 mJ, $p < 0.001$; MPA: 2.8 ± 1.1 mJ vs. 4.2 ± 1.9 mJ, $p = 0.002$; PA: 4.0 ± 1.7 mJ vs. 7.3 ± 3.3 mJ, $p < 0.001$; RPA: 0.7 ± 0.5 mJ vs. 2.0 ± 1.2 mJ, $p < 0.001$).

Peak velocity did not differ in any region of the PA (RVOT: 0.71 ± 0.26 m/s vs. 0.82 ± 0.17 m/s, $p = 0.145$; LPA: 0.71 ± 0.27 m/s vs. 0.81 ± 0.25 m/s, $p = 0.210$; MPA: 0.72 ± 0.26 m/s vs. 0.84 ± 0.22 m/s, $p = 0.092$; RPA: 0.69 ± 0.31 m/s vs. 0.81 ± 0.18 m/s, $p = 0.173$). Forward flow was lower in the RPA (26.08 ± 13.12 mL vs. 33.82 ± 12.08 mL, $p = 0.042$) of PHT patients but otherwise did not differ (RVOT: 74.86 ± 37.25 mL vs. 83.11 ± 23.11 mL, $p = 0.413$; LPA: 43.10 ± 27.79 mL vs. 43.15 ± 13.29 mL, $p = 0.995$; MPA: 80.52 ± 40.39 mL vs. 89.27 ± 19.41 mL, $p = 0.408$). Similarly, net flow was also lower in the RPA (25.84 ± 13.35 mL vs. 33.28 ± 11.83 mL, $p = 0.049$) of PHT patients but otherwise did not differ (RVOT: 72.38 ± 36.23 mL vs. 81.06 ± 23.71 mL, $p = 0.380$; LPA: 43.04 ± 27.84 mL vs. 42.58 ± 13.52 mL, $p = 0.949$; MPA: 79.70 ± 40.75 mL vs. 88.25 ± 20.00 mL, $p = 0.424$). Retrograde flow (0.06 ± 0.14 mL vs. 0.57 ± 1.15 mL, $p = 0.017$) was lower in the LPA of PHT patients but otherwise did not differ (RVOT: 2.48 ± 5.89 mL vs. 2.05 ± 2.48 mL, $p = 0.715$; MPA: 0.81 ± 1.01 mL vs. 1.02 ± 1.65 mL, $p = 0.637$; RPA: 0.24 ± 0.65 mL vs. 0.55 ± 0.79 mL, $p = 0.162$). Furthermore, regurgitation fractions were decreased in the LPA (0.33 ± 0.83% vs. 1.57 ± 3.18% $p = 0.043$) and RPA (0.23 ± 0.54% vs. 1.56 ± 1.89%, $p = 0.001$) but did not differ in the RVOT (3.48 ± 5.60% vs. 2.92 ± 4.20%, $p = 0.695$) or MPA (1.78 ± 3.78% vs. 1.32 ± 2.21%, $p = 0.588$).

3.4. Univariate Analysis

Correlational analyses were reported in Table 4. LVESV correlated with LVEF (R = −0.610, $p < 0.001$). RVEDV correlated with LVEF and RVEF (R = −0.454, $p = 0.005$; R = −0.610, $p < 0.001$). A similar correlation was observed for RVESV (R = −0.677, $p < 0.001$; R = −0.864, $p < 0.001$). MPA-AP (R = 0.385, $p = 0.017$) and MPA-RL (R = 0.418, $p = 0.009$) are significantly correlated to RVEF but not LVEF (MPA-AP: R = 0.097, $p = 0.520$; MPA-RL: R = 0.079; $p = 0.603$). All peak velocity measurements were significantly correlated with LVEF (RVOT: R = 0.451, $p = 0.002$; MPA: R = 0.489, $p = 0.001$; LPA: R = 0.454, $p = 0.002$; RPA: R = 0.527, $p < 0.001$) and RVEF (RVOT: R = 0.416, $p = 0.009$; MPA: R = 0.373, $p = 0.021$; LPA: R = 0.401, $p = 0.013$; RPA: R = 0.420, $p = 0.009$). The MPA RF (R = −0.447, $p = 0.002$), MPA max WSS (R = 0.519, $p < 0.001$) and MPA mean WSS (R = 0.393, $p = 0.007$) correlated with LVEF. LPA mean WSS also correlated with LVEF (R = 0.306, $p = 0.038$). MPA mean Vortex (R = −0.335, $p = 0.023$) and MPA max KE (R = 0.332, $p = 0.024$) also showed a correlation with LVEF.

Table 4. Univariate analysis for ejection fraction.

	LVEF		RVEF	
	R	*p*-Value	R	*p*-Value
Age	0.180	0.232	−0.145	0.384
Sex	0.210	0.161	0.291	0.076
Height	−0.246	0.099	−0.258	0.118
Weight	−0.001	0.997	−0.290	0.078
BSA	−0.058	0.700	−0.307	0.061
Heart Rate	−0.290	0.056	−0.002	0.990
SBP	0.075	0.665	−0.246	0.168
DBP	−0.267	0.116	−0.126	0.486
LVEDV	−0.277	0.062	−0.237	0.152
LVESV	−0.610	<0.001	−0.123	0.463
RVEDV	−0.454	0.005	−0.610	<0.001
RVESV	−0.677	<0.001	−0.864	<0.001
MPA-AP	0.097	0.520	0.385	0.017
MPA-RL	0.079	0.603	0.418	0.009
RPA-AP	−0.001	0.993	0.132	0.429
RPA-CC	−0.0179	0.234	0.081	0.629
LPA-AP	0.061	0.688	0.069	0.679
LPA-CC	0.003	0.982	0.168	0.312
MPA-distance	0.184	0.221	0.082	0.626
RPA-distance	0.024	0.874	−0.037	0.825
LPA-distance	−0.072	0.635	−0.136	0.416
Bifurcation Angle	−0.011	0.944	−0.052	0.755
RVOT Peak Velocity	0.451	0.002	0.416	0.009
MPA Peak Velocity	0.489	0.001	0.373	0.021
RPA Peak Velocity	0.527	<0.001	0.42	0.009
LPA Peak Velocity	0.454	0.002	0.401	0.013
MPA Net Flow	0.196	0.193	0.017	0.917
RPA Net Flow	0.267	0.073	−0.031	0.854
LPA Net Flow	0.100	0.508	−0.012	0.944
MPA Retrograde Flow	0.067	0.660	0.137	0.411
RPA Retrograde Flow	0.281	0.059	0.169	0.311
LPA Retrograde Flow	−0.225	0.132	0.034	0.839
MPA RF	−0.447	0.002	−0.298	0.069
RPA RF	0.059	0.700	−0.010	0.0954
LPA RF	0.158	0.296	−0.094	0.575
MPA Forward Flow	0.195	0.194	0.011	0.950
RPA Forward Flow	0.247	0.098	−0.041	0.807
LPA Forward Flow	0.112	0.458	−0.014	0.936
Max EL	0.261	0.080	−0.006	0.971
Mean EL	0.209	0.164	−0.012	0.943
MPA Max WSS	0.519	<0.001	0.231	0.164
MPA Mean WSS	0.393	0.007	0.281	0.088
RPA Max WSS	0.170	0.260	−0.037	0.826
RPA Mean WSS	0.190	0.206	−0.026	0.876
LPA Max WSS	0.249	0.095	0.036	0.832
LPA Mean WSS	0.306	0.038	0.076	0.651
MPA Max Vortex	−0.267	0.073	−0.168	0.313
MPA Mean Vortex	−0.335	0.023	−0.203	0.221
RPA Max Vortex	−0.107	0.478	−0.229	0.167
RPA Mean Vortex	−0.122	0.420	−0.222	0.180
LPA Max Vortex	0.016	0.916	−0.017	0.918
LPA Mean Vortex	−0.017	0.910	0.011	0.946
MPA Max KE	0.332	0.024	0.097	0.564
RPA Max KE	0.241	0.106	0.004	0.980
LPA Max KE	0.279	0.060	0.054	0.749

Data are expressed as mean ± SD. BSA: body surface area. SBP: systolic blood pressure. DBP: diastolic blood pressure. LVEDV: left ventricle end-diastolic volume. LVESV: left ventricle end-systolic volume. LVEF: left ventricle ejection fraction. RVEDV: right ventricle end-diastolic volume. RVESV: right ventricle end-systolic volume. RVEF: right ventricle ejection fraction. RVOT: right ventricular outflow track. LPA: left pulmonary artery. MPA: main PA. RPA: right PA. AP: anterior–posterior. CC: cranio-caudal. RL: right Left. RF: regurgitant fraction. EL: energy loss. WSS: wall shear stress. KE: kinetic energy.

3.5. Sex Matching for Women

Sex matching analysis is reported in Table 5. When matching a women subset of controls and patients (n = 10), significant differences ($p < 0.05$) were identified for weight; for MPA-AP, MPA-RL, RPA-AP diameters; and for RPA distance. Similarly, advanced parameters showed differences for mean EL MPA (p = 0.025), RPA max WSS (p = 0.003), LPA max WSS (p = 0.019), MPA max vortex (p = 0.003), MPA mean vortex (p = 0.001), RPA mean vortex (p = 0.037), LPA max vortex (p = 0.017), LPA mean vortex (p = 0.001), and RPA max KE (p = 0.004).

Table 5. Matching analysis for women.

	Pulmonary Hypertension (n = 10)	Controls (n = 10)	p-Value
Age (years)	54 ± 17	39 ± 17	0.063
Height (m)	1.59 ± 0.12	1.63 ± 0.04	0. 503
BSA (m^2)	1.85 ± 0.30	1.59 ± 0.26	0.050
Weight (kg)	79 ± 21	61 ± 10	0.027
Heart Rate (bpm)	79 ± 19	67 ± 14	0.124
Systolic Blood Pressure (mmHg)	116 ± 18	104 ± 16	0.171
Diastolic Blood Pressure (mmHg)	63 ± 11	60 ± 13	0.596
LVEF (%)	64 ± 7	60 ± 7	0.208
LVEDV (ml)	118 ± 32	128 ± 24	0.437
LVESV (ml)	48 ± 20	61 ± 32	0.317
RVEF (%)	66 ± 33	61 ± 4	0.733
RVEDV (ml)	163 ± 74	122 ± 35	0.233
RVESV (ml)	75 ± 49	49 ± 14	0.197
MPA-AP (mm)	35 ± 5	28 ± 3	0.002
MPA-RL (mm)	34 ± 7	27 ± 2	0.007
RPA-AP (mm)	21 ± 4	17 ± 3	0.025
RPA-CC (mm)	22 ± 4	20 ± 3	0.123
LPA-AP (mm)	20 ± 5	18 ± 3	0.349
LPA-CC (mm)	21 ± 4	18 ± 2	0.087
MPA-distance (mm)	33 ± 5	31 ± 7	0.455
RPA-distance (mm)	34 ± 6	27 ± 2	0.001
LPA-distance (mm)	33 ± 7	28 ± 7	0.079
Bifurcation Angle (deg)	91 ± 10	97 ± 10	0.187
Max EL MPA (μW)	2.8 ± 1.0	3.1 ± 1.0	0.517
Mean EL MPA (μW)	1.4 ± 0.5	1.0 ± 0.1	0.025
MPA Max WSS (Pa)	1.02 ± 0.21	1.02 ± 0.15	0.962
MPA Mean WSS (Pa)	0.15 ± 0.02	0.16 ± 0.02	0.473
RPA Max WSS (Pa)	0.65 ± 0.26	1.09 ± 0.33	0.003
RPA Mean WSS (Pa)	0.13 ± 0.04	0.16 ± 0.04	0.059
LPA Max WSS (Pa)	0.77 ± 0.30	1.09 ± 0.27	0.019
LPA Mean WSS (Pa)	0.15 ± 0.05	0.17 ± 0.03	0.296
MPA Max Vortex (mm^3)	28.40 ± 9.73	16.04 ± 5.53	0.003
MPA Mean Vortex (mm^3)	20.25 ± 8.49	9.15 ± 3.38	0.001
RPA Max Vortex (mm^3)	3.95 ± 1.95	2.78 ± 1.81	0.181
RPA Mean Vortex (mm^3)	2.60 ± 1.35	1.46 ± 0.86	0.037
LPA Max Vortex (mm^3)	6.50 ± 3.08	3.79 ± 1.07	0.017
LPA Mean Vortex (mm^3)	4.33 ± 2.16	1.65 ± 0.53	0.001
MPA Max KE (mJ)	2.83 ± 0.93	3.20 ± 1.0	0.413
RPA Max KE (mJ)	0.67 ± 0.03	1.44 ± 0.67	0.004
LPA Max KE (mJ)	1.17 ± 0.66	1.74 ± 0.94	0.131

Data are expressed as mean ± SD. BSA: body surface area. SBP: systolic blood pressure. DBP: diastolic blood pressure. LVEDV: left ventricle end-diastolic volume. LVESV: left ventricle end-systolic volume. LVEF: left ventricle ejection fraction. RVEDV: right ventricle end-diastolic volume. RVESV: right ventricle end-systolic volume. RVEF: right ventricle ejection fraction. RVOT: right ventricular outflow track. LPA: left pulmonary artery. MPA: main PA. RPA: right PA. AP: anterior–posterior. CC: cranio-caudal. RL: right Left. RF: regurgitant fraction. EL: energy loss. WSS: wall shear stress. KE: kinetic energy.

4. Discussion

In this study, we demonstrate that PHT is associated with both hemodynamic and anatomic changes throughout the PA. Patients with PHT demonstrated multiple changes across a variety of hemodynamic parameters, including EL, WSS, vorticity, flow direction and flow velocity. In addition, pulmonary artery dilation was observed alongside changes in bifurcation geometry. Sex matching for women identified differences for several anatomical and advanced hemodynamic biomarkers. Taken together, these findings provide additional evidence that hemodynamic and anatomic biomarkers possess clinical relevance for diagnosis of the disease.

Although hemodynamic changes are emerging biomarkers of the disease, they remain poorly understood. Previous research suggests patients with PHT demonstrate lower wall shear stress [5,7]. Although these studies were conducted using an adult PHT population, the findings possess relevance to the broader definition of PHT. For vorticity, a recent study conducted by Kroeger et al. found that vortices are observed in healthy controls and patients with PHT [19]. Similarly, Schafer et al. identified reduced vorticity but increased helicity in patients with PHT [6]. For EL, Han et al. demonstrated greater total energy losses over the cardiac cycle throughout the entirety of the PA [20] in PHT patients. PHT patients may also demonstrate greater retrograde flow and lower peak velocity [21,22]. In contrast, anatomical changes are more well-studied using CT, identifying dilation of the main pulmonary artery in various stages of the disease [23,24]. These findings have been replicated using MRI and appear to be prevalent in the later stages [25].

Our findings generally corroborate results of previous studies and provide insight into novel biomarkers. We observed lower WSS, maximum energy attained, and regurgitation fractions in the LPA and RPA in patients with PHT. Maximum EL at peak systole remained lower in patients than in controls in all PA sections. Mean EL tended to be also lower, except for the LPA. EL reported in our study was quantified at peak systole, which provides further support in addition of previous findings reported over the cardiac cycle. Contrary to other studies, we found that vorticity tended to increase in PHT patients. For anatomical changes, we observed dilation in the MPA and part of the RPA. Of note, we are the first study to our knowledge to identify greater bifurcation distances associated with PHT. We speculate this could be an additional form of anatomical remodeling associated with disease progression. However, the exact significance is unknown. Further studies investigating changes in bifurcation geometry at various disease stages are needed to better understand their significance. The difference between the RPA and LPA characterization may be influenced by the patient's anatomy and the hemodynamic resistance from the lungs. However, the latter requires additional investigation. Furthermore, sex-matching analysis confirmed the impact of advanced hemodynamic parameters in PHT patients and highlighted the importance of subject matching in future studies.

This study has a few limitations. First, our study was limited by the spatiotemporal resolution constraints of MRI. For example, while we could visualize larger vortex networks, smaller vortices occurring at faster speeds exceeding the available resolution could be missed. Second, as a single centre study, the addition of multiple centres would improve the generalization of results. Furthermore, due to its retrospective nature, the sample size and power is limited. Third, our control patients were not age and sex-normalized to the patient cohort, which suggests normal cardiac aging and sex differences could have impacted results. Fourth, PHT is a progressive multi-factorial condition with diverse underlying etiology. Clinical protocols did not mandate further characterization of the disease into smaller subgroups such as pulmonary arterial hypertension, or patients at varying stages of the disease, limiting the characterization of disease progression. The relatively small cohort size did not allow one to explore the effect of medical therapy in 4D-flow derived parameters. The latter remains an important aspect to assess in future studies. Finally, we did not explore the right ventriculo-pulmonary arterial coupling in the present cohort. This coupling analysis may produce a better understanding of the bidirectional effects between the RV and the pulmonary artery tree.

5. Conclusions

Adult PHT patients demonstrate hemodynamic and anatomic differences throughout the PA. In addition to providing further support for established disease phenotypes such as reduced WSS and PA dilatation, these findings suggest that novel biomarkers such as energy changes and changes in bifurcation geometry may possess clinical relevance for PHT patients. Future studies are needed to investigate these phenomena at varying stages of the disease and determine how they may be affected by underlying etiology.

Author Contributions: Conceptualization, Z.M.H. and J.G.; methodology, J.G.; software, J.G.; validation, Z.M.H. and J.G.; formal analysis, Z.M.H. and J.G.; investigation, Z.M.H. and J.G.; resources, J.G.; data curation, Z.M.H. and J.G.; writing—original draft preparation, Z.M.H. and J.G.; writing—review and editing, J.G.; visualization, Z.M.H. and J.G.; supervision, J.G.; project administration, J.G.; funding acquisition, J.G. All authors have read and agreed to the published version of the manuscript.

Funding: This research was funded by The University of Calgary, URGC SEM #1054341; J.G. start-up funding. We acknowledge the support of the Natural Science and Engineering Research Council of Canada/Conseil de recherche en science naturelles et en génie du Canada, RGPIN-2020-04549 and DGECR-2020-00204.

Institutional Review Board Statement: The study was conducted according to the guidelines of the Declaration of Helsinki and approved by the Conjoint Health Research Ethics Board of University of Calgary (REB13-0902 approved on 6/18/2014 and currently active.

Informed Consent Statement: Written informed consent was obtained from all subjects involved in the study.

Data Availability Statement: The anonymized data presented in this study are available on request from the corresponding author. The data are not publicity available due to privacy and ethical restrictions.

Conflicts of Interest: The authors declare no conflict of interest.

References

1. Galiè, N.; Humbert, M.; Vachiery, J.-L.; Gibbs, S.; Lang, I.; Torbicki, A.; Simonneau, G.; Peacock, A.; Vonk Noordegraaf, A.; Beghetti, M.; et al. 2015 ESC/ERS Guidelines for the diagnosis and treatment of pulmonary hypertension: The Joint Task Force for the Diagnosis and Treatment of Pulmonary Hypertension of the European Society of Cardiology (ESC) and the European Respiratory Society (ERS): Endor. *Eur. Heart J.* **2016**, *37*, 67–119. [CrossRef]
2. Deng, Y.; Guo, S.-L.; Wu, W.-F.; Wang, Q.; Su, H.-Y.; Tan, Z.; Wang, F.; He, Q.-Y. Right atrial evaluation in patients with pulmonary hypertension. *J. Ultrasound Med.* **2016**, *35*, 49–61. [CrossRef]
3. Freed, B.H.; Collins, J.D.; Francois, C.J.; Barker, A.J.; Cuttica, M.J.; Chesler, N.C.; Markl, M.; Shah, S.J. MR and CT imaging for the evaluation of pulmonary hypertension. *JACC Cardiovasc. Imaging* **2016**, *9*, 715–732. [CrossRef]
4. Nordmeyer, S.; Riesenkampff, E.; Crelier, G.; Khasheei, A.; Schnackenburg, B.; Berger, F.; Kuehne, T. Flow-sensitive four-dimensional cine magnetic resonance imaging for offline blood flow quantification in multiple vessels: A validation study. *J. Magn. Reson. Imaging* **2010**, *32*, 677–683. [CrossRef] [PubMed]
5. Barker, A.J.; Roldán-Alzate, A.; Entezari, P.; Shah, S.J.; Chesler, N.C.; Wieben, O.; Markl, M.; Francois, C.J. Four-dimensional flow assessment of pulmonary artery flow and wall shear stress in adult pulmonary arterial hypertension: Results from two institutions. *Magn. Reson. Med.* **2015**, *73*, 1904–1913. [CrossRef]
6. Schäfer, M.; Barker, A.J.; Kheyfets, V.; Stenmark, K.R.; Crapo, J.; Yeager, M.E.; Truong, U.; Buckner, J.K.; Fenster, B.E.; Hunter, K.S. Helicity and vorticity of pulmonary arterial flow in patients with pulmonary hypertension: Quantitative analysis of flow formations. *J. Am. Hear. Assoc.* **2017**, *6*. [CrossRef]
7. Tang, B.T.; Pickard, S.S.; Chan, F.P.; Tsao, P.S.; Taylor, C.A.; Feinstein, J.A. Wall shear stress is decreased in the pulmonary arteries of patients with pulmonary arterial hypertension: An image-based, computational fluid dynamics study. *Pulm. Circ.* **2012**, *2*, 470–476. [CrossRef]
8. Boerrigter, B.; Mauritz, G.-J.; Marcus, J.T.; Helderman, F.; Postmus, P.; Westerhof, N.; Vonk-Noordegraaf, A. Progressive dilatation of the main pulmonary artery is a characteristic of pulmonary arterial hypertension and is not related to changes in pressure. *Chest* **2010**, *138*, 1395–1401. [CrossRef]
9. Kramer, C.M.; Barkhausen, J.; Bucciarelli-Ducci, C.; Flamm, S.D.; Kim, R.J.; Nagel, E. Standardized cardiovascular magnetic resonance imaging (CMR) protocols: 2020 update. *J. Cardiovasc. Magn. Reson.* **2020**, *22*, 17. [CrossRef]
10. Garcia, J.; Sheitt, H.; Bristow, M.S.; Lydell, C.; Howarth, A.G.; Heydari, B.; Prato, F.S.; Drangova, M.; Thornhill, R.E.; Nery, P.; et al. Left atrial vortex size and velocity distributions by 4D flow MRI in patients with paroxysmal atrial fibrillation: Associations with age and CHA 2 DS 2 -VASc risk score. *J. Magn. Reson. Imaging* **2020**, *51*, 871–884. [CrossRef]
11. Garcia, J.; Beckie, K.; Hassanabad, A.F.; Sojoudi, A.; White, J.A. Aortic and mitral flow quantification using dynamic valve tracking and machine learning: Prospective study assessing static and dynamic plane repeatability, variability and agreement. *JRSM Cardiovasc. Dis.* **2021**, *10*. [CrossRef]
12. Geeraert, P.; Jamalidinan, F.; Hassanabad, A.F.; Sojoudi, A.; Bristow, M.; Lydell, C.; Fedak, P.W.M.; White, J.A.; Garcia, J. Bicuspid aortic valve disease is associated with abnormal wall shear stress, viscous energy loss, and pressure drop within the ascending thoracic aorta. *Medicine* **2021**, *100*, e26518. [CrossRef] [PubMed]

13. Geeraert, P.; Jamalidinan, F.; Burns, F.; Jarvis, K.; Bristow, M.S.; Lydell, C.; Tobon, S.S.H.; Alonso, B.d.C.; Fedak, P.W.M.; White, J.A.; et al. Hemodynamic assessment in bicuspid aortic valve disease and aortic dilation: New insights from voxel-by-voxel analysis of reverse flow, stasis, and energetics. *Front. Bioeng. Biotechnol.* **2022**, *9*, 725113. [CrossRef] [PubMed]
14. Hassanabad, A.F.; Burns, F.; Bristow, M.S.; Lydell, C.; Howarth, A.G.; Heydari, B.; Gao, X.; Fedak, P.W.; White, J.A.; Garcia, J. Pressure drop mapping using 4D flow MRI in patients with bicuspid aortic valve disease: A novel marker of valvular obstruction. *Magn. Reson. Imaging* **2020**, *65*, 175–182. [CrossRef]
15. Rose, M.J.; Jarvis, K.; Chowdhary, V.; Barker, A.J.; Allen, B.D.; Robinson, J.D.; Markl, M.; Rigsby, C.; Schnell, S. Efficient method for volumetric assessment of peak blood flow velocity using 4D flow MRI. *J. Magn. Reson. Imaging* **2016**, *44*, 1673–1682. [CrossRef] [PubMed]
16. Ooij, P.v.; Garcia, J.; Potters, W.V.; Malaisrie, S.C.; Collins, J.D.; Carr, J.C.; Markl, M.; Barker, A.J. Age-related changes in aortic 3D blood flow velocities and wall shear stress: Implications for the identification of altered hemodynamics in patients with aortic valve disease. *J. Magn. Reson. Imaging* **2016**, *43*, 1239–1249. [CrossRef] [PubMed]
17. Potters, W.V.; Ooij, P.v.; Marquering, H.; vanBavel, E.; Nederveen, A.J. Volumetric arterial wall shear stress calculation based on cine phase contrast MRI. *J. Magn. Reson. Imaging* **2015**, *41*, 505–516. [CrossRef] [PubMed]
18. Barker, A.J.; Ooij, P.v.; Bandi, K.; Garcia, J.; Albaghdadi, M.; McCarthy, P.; Bonow, R.O.; Carr, J.; Collins, J.; Malaisrie, S.C.; et al. Viscous energy loss in the presence of abnormal aortic flow. *Magn. Reson. Med.* **2014**, *72*, 620–628. [CrossRef]
19. Kroeger, J.R.; Stackl, M.; Weiss, K.; Baeßler, B.; Gerhardt, F.; Rosenkranz, S.; Maintz, D.; Giese, D.; Bunck, A.C. K-t accelerated multi-VENC 4D flow MRI improves vortex assessment in pulmonary hypertension. *Eur. J. Radiol.* **2021**, *145*, 110035. [CrossRef]
20. Han, Q.J.; Witsche, W.R.T.; Fang-Yen, C.M.; Arkles, J.S.; Barker, A.J.; Forfia, P.R.; Han, Y. Altered right ventricular kinetic energy work density and viscous energy dissipation in patients with pulmonary arterial hypertension: A pilot study using 4D flow MRI. *PLoS ONE* **2015**, *10*, e0138365. [CrossRef]
21. Helderman, F.; Maurit, G.-J.; Andringa, K.E.; Vonk-Noordegraaf, A.; Marcus, J.T. Early onset of retrograde flow in the main pulmonary artery is a characteristic of pulmonary arterial hypertension. *J. Magn. Reson. Imaging* **2011**, *33*, 1362–1368. [CrossRef] [PubMed]
22. Kondo, C.; Caputo, G.R.; Masui, T.; Foster, E.; O'Sullivan, M.; Stulbarg, M.S.; Golden, J.; Catterjee, K.; Higgins, C.B. Pulmonary hypertension: Pulmonary flow quantification and flow profile analysis with velocity-encoded cine MR imaging. *Radiology* **1992**, *183*, 751–758. [CrossRef]
23. Burger, I.A.; Husmann, L.; Herzog, B.A.; Buechel, R.R.; Pazhenkottil, A.P.; Ghadri, J.R.; Nkoulou, R.N.; Jenni, R.; Russi, E.W.; Kaufmann, P.A. Main pulmonary artery diameter from attenuation correction CT scans in cardiac SPECT accurately predicts pulmonary hypertension. *J. Nucl. Cardiol.* **2011**, *18*, 634–641. [CrossRef]
24. Lange, T.J.; Bornia, C.; Stiefel, J.L.; Stroszczynski, C.; Arzt, M.; Pfeifer, M.; Hamer, O.W. Increased pulmonary artery diameter on chest computed tomography can predict borderline pulmonary hypertension. *Pulm. Circ.* **2013**, *3*, 363–368. [CrossRef]
25. Badagliacca, R.; Poscia, R.; Pezzuto, B.; Papa, S.; Nona, A.; Mancone, M.; Mezzapesa, M.; Nocioni, M.; Sciomer, S.; Valli, G.; et al. Pulmonary arterial dilatation in pulmonary hypertension: Prevalence and prognostic relevance. *Cardiology* **2012**, *121*, 76–82. [CrossRef] [PubMed]

applied sciences

MDPI

Article

Impact of Respiratory Gating on Hemodynamic Parameters from 4D Flow MRI

Esteban Denecken [1,2,3,4], Julio Sotelo [1,3,4,5], Cristobal Arrieta [1,3], Marcelo E. Andia [1,3,4,6] and Sergio Uribe [1,3,4,6,*]

1 Biomedical Imaging Center, Pontificia Universidad Católica de Chile, Santiago 782 0436, Chile; ejdenecken@uc.cl (E.D.); julio.sotelo@uv.cl (J.S.); ciarriet@uc.cl (C.A.); meandia@uc.cl (M.E.A.)
2 Department of Electrical Engineering, Pontificia Universidad Católica de Chile, Santiago 782 0436, Chile
3 ANID, Millennium Science Initiative Program, Millennium Nucleus for Cardiovascular Magnetic Resonance, Santiago 782 0436, Chile
4 Instituto Milenio Intelligent Healthcare Engineering, Santiago 782 0436, Chile
5 School of Biomedical Engineering, Universidad de Valparaíso, Valparaiso 236 2905, Chile
6 Department of Radiology, School of Medicine, Pontificia Universidad Católica de Chile, Santiago 832 0000, Chile
* Correspondence: suribe@uc.cl

Abstract: The hemodynamic parameters from 4D flow datasets have shown promising diagnostic value in different cardiovascular pathologies. However, the behavior of these parameters can be affected when the 4D flow data are corrupted by respiratory motion. The purpose of this work was to perform a quantitative comparison between hemodynamic parameters computed from 4D flow cardiac MRI both with and without respiratory self-gating. We considered 4D flow MRI data from 15 healthy volunteers (10 men and 5 women, 30.40 ± 6.23 years of age) that were acquired at 3T. Using a semiautomatic segmentation process of the aorta, we obtained the hemodynamic parameters from the 4D flow MRI, with and without respiratory self-gating. A statistical analysis, using the Wilcoxon signed-rank test and Bland–Altman, was performed to compare the hemodynamic parameters from both acquisitions. We found that the calculations of the hemodynamic parameters from 4D flow data that were acquired without respiratory self-gating showed underestimated values in the aortic arch, and the descending and diaphragmatic aorta. We also found a significant variability of the hemodynamic parameters in the ascending aorta of healthy volunteers when comparing both methods. The 4D flow MRI requires respiratory compensation to provide reliable calculations of hemodynamic parameters.

Keywords: hemodynamics; 4D flow; self-gating; aorta

Citation: Denecken, E.; Sotelo, J.; Arrieta, C.; Andia, M.E.; Uribe, S. Impact of Respiratory Gating on Hemodynamic Parameters from 4D Flow MRI. *Appl. Sci.* **2022**, *12*, 2943. https://doi.org/10.3390/app12062943

Academic Editor: Julio Garcia Flores

Received: 15 December 2021
Accepted: 26 January 2022
Published: 14 March 2022

1. Introduction

Several strategies have been proposed to suppress respiratory motion in cardiac MRI. Most of these techniques are used to track the position of the diaphragm and can be combined with a respiratory-based k-space reordering scheme [1–6]. These techniques include pencil beam navigators [1,2], cross-pair excitation to acquire a column of pixels across the lung-to-liver interface, self-gating strategies based on k_0 points [3], k_0 profiles [4] or 2D image navigators [5].

Newer approaches have achieved respiratory resolved cardiac images with a near 100% respiratory navigator efficiency [7,8]. These approaches have been used to obtain single- or dual-phase 3D whole heart balanced steady state free precession (b-SSFP) images [2], 2D or 3D b-SSFP cine images [9], and 4D flow MRI data [4]. The use of cine b-SSFP respiratory compensated data has remarkably improved the image quality, as well as the anatomical and functional accuracy [7,8]. Nevertheless, there is a debate whether the use of respiratory motion compensation techniques improves the quantification of velocity-derived parameters from the 4D flow MRI [10].

Applying respiratory compensation to 4D flow MRI increases the already long scanning time of this sequence. Moreover, its impact on the accuracy of the hemodynamic parameters remains unclear. Some authors propose that the use of respiratory self-gating techniques has a clear impact on 4D flow measurements by looking at the stroke volume [4,11]. Other studies propose that aorta flow volume quantification does not need respiratory compensation [12]. Other approaches that used profile reordering for respiratory compensation in 4D flow MRI [13,14] suggest that the use of a motion compensation technique increases the accuracy of 4D flow measurements. Most of this discussion has been based on the net flow going through a vessel of interest. However, the net flow is a parameter that averages the velocities in a region of interest, which may hide the distortion of respiratory motion on the velocity field.

Over recent years, several parameters from 4D flow datasets have been derived, such as Wall Shear Stress (WSS), vorticity, helicity, kinetic energy, and viscous dissipation, among others [10,15]. These parameters have been applied in several groups of patients and have shown promising diagnostic value in different cardiovascular pathologies. Nevertheless, these parameters need to be carefully used and compared since their value depends on the resolution and the level of noise of the acquisitions [16]. We also know little about the behavior of these novel parameters when the 4D flow data are corrupted by respiratory motion.

Some of these new hemodynamic parameters are directly proportional to velocity, but others are related to spatial derivatives. Therefore, any rapid perturbation of the velocity field that is introduced by noise or motion could impact the precision and accuracy of the computations. We hypothesized that there are significant differences in the hemodynamic parameters obtained from 4D flow MRI that are acquired with and without respiratory gating in healthy volunteers. The purpose of this work was to perform a quantitative comparison of hemodynamic parameters obtained from 4D flow MRI data that were acquired both with and without respiratory self-gating. We compared the differences between the following hemodynamic parameters, computed in the thoracic aorta: velocity; WSS; WSS axial (WSSA) and circumferential (WSSC) components; vorticity; viscous dissipation; kinetic energy; and energy loss. We also considered the effect of the segmentation process, comparing the segmentations from respiratory self-gated and non-gated data using the DICE as a global index and computing mesh distances to study the local differences.

2. Materials and Methods

2.1. Data Acquisition

This study included thoracic aorta 4D flow MRI data from 15 healthy volunteers (10 men and 5 women, 30.40 $\pm$ 6.23 years of age), which had been previously acquired. The study was conducted according to the guidelines of the Declaration of Helsinki and was approved by the Ethics Committee at Pontificia Universidad Católica (protocol code 171,127,001, 20 June 2019). A 3T Philips MRI scanner was used to acquire the data, with and without respiratory self-gating. The respiratory self-gated strategy consists of the continuous acquisition of the k0 profile and the cross-correlation to determine the respiratory positions [4]. A window of 7 mm was used to collect the data during expiration [4]. A summary of the acquisition parameters is shown in Table 1. The non-respiratory gating scan was acquired with two repetitions (NSA, NEX = 2) in order to have a similar acquisition time to the respiratory self-gated scan.

2.2. Segmentation and Data Processing

Both 4D Flow MRI datasets (with and without respiratory self-gating) were post-processed using an in-house MATLAB toolbox (MathWorks, Natick, MA, USA) [17]. The image processing steps involved the segmentation of the thoracic aorta and the generation of a tetrahedral finite element mesh (characteristic length 1 mm), as described in [18]. The velocity, WSS, WSSA, WSSC, vorticity, viscous dissipation, kinetic energy, and energy loss were obtained from the thoracic aorta 4D flow images, using a least square projection method based on the finite elements described in [18–20]. All parameters were obtained at a representative peak systolic cardiac phase. These parameters have shown potential value in diagnostics [21]. Angiographic 4D flow images were segmented using a semiautomatic process based on the thresholding, labeling, and manual separation of the vessels. The analysis was performed at the peak systolic phase.

The computations of the hemodynamic parameters were performed using the in-house MATLAB toolbox developed by Sotelo et al. [17] and were exhaustively validated [18–20]. The velocity values were transferred to each node of the mesh from the 3D PC-MRI datasets using a cubic interpolation [18].

Table 1. The MRI acquisition parameters.

Parameter		
Field of view, mm	236 × 226 × 134	
Voxel size, mm	2.34 × 2.34 × 2.5	
Cardiac phases	25	
VENC, cm/s	250	
Echo time, ms	2.66	
Repetition time, ms	4.78–4.80	
Flip angle	5°	
	Self-Gated	Non-Gated
Temporal resolution, ms	37.62 ± 4.43	36.11 ± 4.18
Heart rate, bpm	66.33 ± 7.62	67.93 ± 8.03
Acquisition time, min	15.00 ± 2.86	12.33 ± 0.37

2.3. Statistical Analysis

We assessed the differences between the hemodynamic parameters that were obtained with and without respiratory self-gating over four different regions of the aorta: region I = ascending aorta (between the sinus of Valsalva and before the brachiocephalic trunk); region II = aortic arch (before the brachiocephalic trunk and isthmus level); region III = superior descending aorta (between the isthmus level and the Valsalva level); and region IV = posterior descending aorta (between the Valsalva level and the diaphragmatic aorta). We computed the mean of the measurements in each region to perform the statistical analysis that is presented herein.

Furthermore, we calculated the DICE coefficient between the segmentations of the data that were obtained with and without respiratory self-gating and the local differences between the mesh surfaces of each section. The local differences between both meshes were calculated using the Euclidean distance, as reported in [22].

We analyzed the mean, standard deviation, and coefficient of variation (CV) of each hemodynamic parameter. The Wilcoxon signed-rank test and Bland–Altman plots were also performed between the respiratory self-gated and non-gated data, using a statistical significance of 0.05. We computed the Wilcoxon signed-rank test using the SPSS statistical analysis software and the Bland–Altman plots using MATLAB.

3. Results

Figure 1 shows a representative case of the hemodynamic parameters in the aorta that was obtained from respiratory and non-respiratory gated 4D flow MRI data. By visual inspection, the images look similar. Nevertheless, differences can be observed in the WSS, viscous dissipation, energy loss, and kinetic energy.

Table 2 shows the mean and CV obtained for each hemodynamic parameter at each section of the aorta. The mean of the hemodynamic parameters was similar in section I, but it showed differences in the other sections. Further, the CV values were greater in the 4D flow data acquired without respiratory self-gating compared to the self-gated data. The CV values were greater in section I.

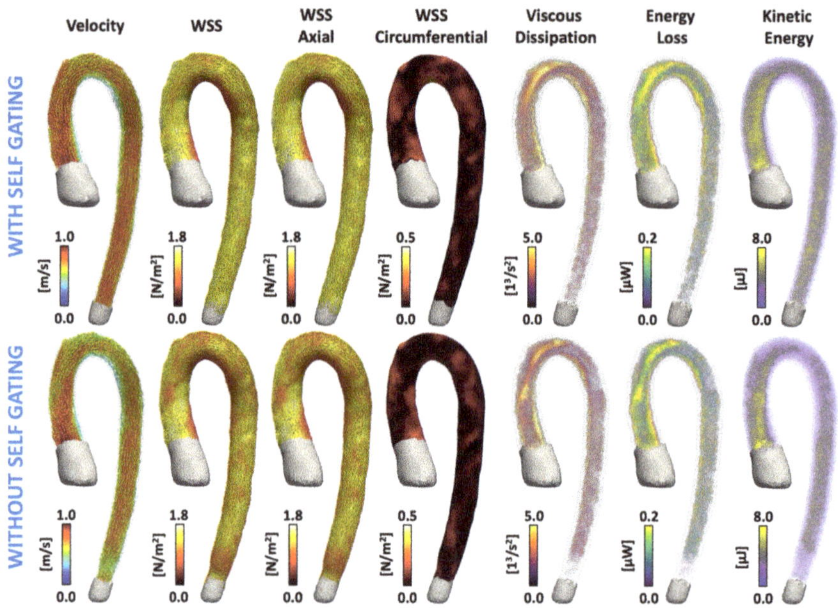

Figure 1. The hemodynamic parameters obtained from 4D flow MRI, both with and without self-gating. In this figure, we show visual representations of the hemodynamic parameters for one volunteer, which were measured at a representative peak systolic cardiac phase.

Table 2. The hemodynamic parameter means and the coefficient of variation (CV) of the mean values at each section of the vessel for the self-gated and non-gated data, which were measured at a representative peak systolic cardiac phase.

	Section I		Section II	
Hemodynamic Parameter	**Self-Gated Mean (CV)**	**Non-Gated Mean (CV)**	**Self-Gated Mean (CV)**	**Non-Gated Mean (CV)**
Velocity (m/s)	0.686 (0.197)	0.652 (0.251)	0.571 (0.159)	0.527 (0.204)
WSS (N/m^2)	1.320 (0.186)	1.265 (0.245)	1.165 (0.138)	1.092 (0.209)
WSSA (N/m^2)	1.308 (0.187)	1.252 (0.242)	1.157 (0.135)	1.084 (0.208)
WSSC (N/m^2)	0.121 (0.343)	0.122 (0.505)	0.101 (0.333)	0.094 (0.302)
Vorticity (1/s)	45.549 (0.310)	45.838 (0.330)	42.008 (0.345)	39.730 (0.370)
Viscous Dissipation (1/s^2)	4.569 (0.394)	4.826 (0.590)	3.047 (0.377)	2.611 (0.370)
Kinetic Energy (μJ)	0.258 (0.404)	0.271 (0.601)	0.172 (0.386)	0.147 (0.384)
Energy Loss (μW)	4.303 (0.392)	4.002 (0.479)	2.976 (0.324)	2.572 (0.401)
	Section III		**Section IV**	
Hemodynamic Parameter	**Self-Gated Mean (CV)**	**Non-Gated Mean (CV)**	**Self-Gated Mean (CV)**	**Non-Gated Mean (CV)**
Velocity (m/s)	0.558 (0.205)	0.501 (0.318)	0.528 (0.277)	0.445 (0.375)
WSS (N/m^2)	1.230 (0.198)	1.099 (0.318)	1.224 (0.268)	1.027 (0.384)
WSSA (N/m^2)	1.228 (0.198)	1.096 (0.319)	1.222 (0.268)	1.024 (0.385)
WSSC (N/m^2)	0.048 (0.314)	0.049 (0.316)	0.047 (0.254)	0.051 (0.330)
Vorticity (1/s)	35.541 (0.258)	32.558 (0.306)	41.900 (0.312)	34.369 (0.287)
Viscous Dissipation (1/s^2)	1.999 (0.471)	1.753 (0.580)	2.394 (0.643)	1.803 (0.545)
Kinetic Energy (μJ)	0.112 (0.477)	0.097 (0.588)	0.132 (0.637)	0.099 (0.553)
Energy Loss (μW)	2.995 (0.410)	2.523 (0.579)	2.873 (0.510)	2.184 (0.642)

Table 3 shows the results obtained from the statistical analysis. The Wilcoxon signed-rank test showed significant differences between respiratory self-gating and non-respiratory gating in sections II, III, and IV. Differences in velocity, WSS, and WSSA were observed in sections II, III, and IV. There were also statistical differences for viscous dissipation, kinetic energy, and energy loss at sections II and IV.

Table 3. The Wilcoxon signed-rank test *p*-values for the mean values of the hemodynamic parameters at each section of the vessel, between the self-gated and non-gated data.

Hemodynamic Parameter	Section I	Section II	Section III	Section IV
Velocity (m/s)	0.1688	0.0125 *	0.0020 *	0.0003 *
WSS (N/m^2)	0.4212	0.0302 *	0.0015 *	0.0002 *
WSSA (N/m^2)	0.3894	0.0302 *	0.0015 *	0.0002 *
WSSC (N/m^2)	0.4543	0.0946	0.6788	0.3303
Vorticity (1/s)	0.6788	0.1354	0.0833	0.0015 *
Viscous Dissipation (1/s^2)	0.6788	0.0215 *	0.0946	0.0302 *
Kinetic Energy (μJ)	0.8904	0.0103 *	0.0554	0.0302 *
Energy Loss (μW)	0.2524	0.0125 *	0.0034 *	0.0003 *

* Existing significative differences.

In Appendix A, Table A1 shows the mean differences and 1.96 times the standard deviation of the Bland–Altman analysis. The 4D flow data with self-gating are used as a reference. The Bland–Altman plots for the velocity and WSS in section I (Figure 2) showed a greater dispersion than sections II, III, and IV. They also show the underestimated values from the data acquired without self-gating. The Bland–Altman plots for velocity, WSS, WSSA, WSSC, vorticity, viscous dissipation, kinetic energy, and energy loss plots are shown in Figure A1. In general, all computations of the hemodynamic parameters acquired without self-gating were underestimated when no respiratory gating was performed. There was a greater difference in velocity, WSS, WSSA, and kinetic energy, and this bias increased along the aortic sections. The standard deviation was greater for most of the parameters in section I compared to other sections.

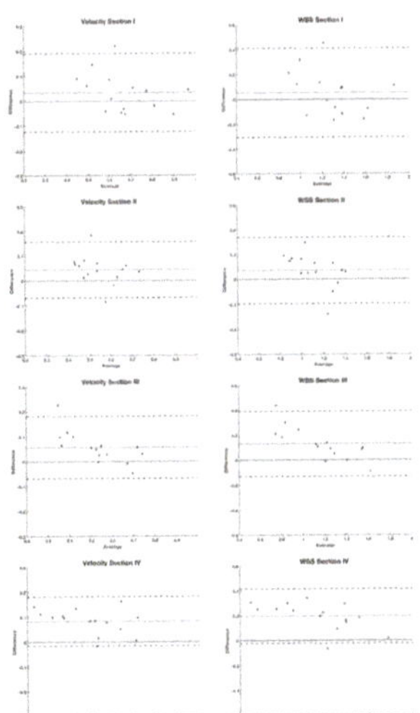

Figure 2. The Bland–Altman plots for velocity and Wall Shear Stress at each section. The difference in Bland–Altman plots was computed as $SG - NG$, where SG is the data with self-gating and NG is the data without self-gating. Appendix A shows the Bland–Altman plots of all hemodynamic parameters considered in this work.

Finally, the mean and range of the DICE coefficient between the segmentation of the self-gated and non-gated data were 0.918 ± 0.056. This result, which considered the whole segmented aorta, showed a high intersection between both segmentations.

A graphical analysis of the local differences [22] between the meshes that were generated from both segmentations shows that the disagreements between them are located at the extremes of the aorta (Figure A2A in Appendix B for one volunteer). The local differences of the segmentations at each section had small variations (Table A2 in Appendix B) and the histogram shows that 88.36% of the differences at each section corresponded to distances lower than or equal to 1 mm (Figure A2B). The local differences were also similarly distributed around the 0% error and thus, there was not a consistent bias between both segmentations.

4. Discussion

In this work, we compared the hemodynamic parameters computed from thoracic aorta 4D flow MRI data that were obtained with and without respiratory self-gating. The statistical descriptors showed that the data obtained with self-gating had less variability than the data obtained without self-gating (Table 2). The Wilcoxon signed-rank test showed significant differences in most of the parameters calculated between both acquisitions at sections II, III, and IV and minimal differences at section I (ascending aorta). Nevertheless, the computations in section I showed a greater variability.

The Bland–Altman analysis showed that the parameters computed without respiratory self-gating were underestimated in comparison to the parameters computed with self-gating and that the underestimation increased along sections II, III, and IV. Section I did not show bias; however, it presented a greater dispersion of the differences for each hemodynamic parameter (see the Bland–Altman plots in Figure A1 in Appendix A). Therefore, for section I, there was not a clear trend of the under- or overestimation of the parameters that were computed. The differences for each volunteer were greater in section I compared to the other sections, which is an important result that is masked by the averaging of all volunteers.

The only parameter that did not show any statistical difference was the WSSC, which is probably due to the near to 0 value of the WSSC in this group of volunteers. An assessment of the impact of respiratory motion on this parameter in a cohort of volunteers or patients with a greater value of WSSC is necessary. As shown by Sotelo et al., there are significant differences in WSSC between healthy volunteers and BAV patients [19].

We also evaluated the impact of the segmentation process as a potential cause of the statistical differences. However, the DICE coefficient and analyses of the local differences at each section showed high agreement between both segmentations and the differences were minimal and unbiased. Therefore, the segmentation did not influence the differences that were observed in the hemodynamic parameters.

A limitation of this study is that we only included healthy volunteers with laminar flow patterns. The next step is to perform a similar study in a group of patients with non-laminar flow patterns or high Reynolds numbers, such as patients with a bicuspid aortic valve (BAV) or with aortic coarctation. Unfortunately, due to the long acquisition time, acquiring two datasets of 4D flow MRI in patients is challenging. Another limitation of the study is that we only assessed the impact of respiratory gating on the hemodynamic parameters at peak systole. The main reason for this was that the poor signal-to-noise ratio of the other cardiac phases makes the segmentation process very challenging. Although it is expected that respiratory motion artifacts may have a stronger effect at lower velocities, this still needs to be proven. Finally, flow parameters as net or regurgitant flow were not included because these results have been presented before using the same datasets [4]. The purpose of this paper was to assess how other hemodynamic parameters are affected by respiratory motion.

In conclusion, we found that the computations of the hemodynamic parameters from 4D flow data using self-gating acquisition showed statistical differences from the computations from acquisitions without respiratory compensation for the aortic arch and the descending and diaphragmatic aorta in healthy volunteers. The Bland–Altman plots support this fact, showing an underestimation of the hemodynamic parameters acquired without any respiratory compensation. Our results suggest that the use of respiratory motion compensation techniques is mandatory for the accurate evaluation of 4D flow hemodynamic parameters.

Author Contributions: Conceptualization, S.U.; methodology, J.S., M.E.A., S.U. and E.D.; software, J.S.; validation, J.S. and M.E.A.; formal analysis, E.D.; resources, S.U.; data curation, E.D.; writing—original draft preparation, E.D.; writing—review and editing, S.U., J.S. and C.A.; visualization, J.S. and E.D.; supervision, S.U.; project administration, S.U.; funding acquisition, S.U. All authors have read and agreed to the published version of the manuscript.

Funding: This work was funded by ANID—Millennium Science Initiative Program—ICN2021_004, ANID-PIA-ACT192064, ANID—Millennium Science Initiative Program—NCN17_129 and FONDECYT # 1181057. Sotelo J. would like to thank CONICYT-FONDECYT Postdoctorado 2017 #3170737 and FONDECYT de Iniciación en Investigación 2020 #11200481. Andia M. would like to thank FONDECYT # 1180525. Arrieta C. was partially funded by CONICYT FONDECYT Postdoctorado 2019 #3190763, CONICYT PCI REDES 180090, and ANID—Millennium Science Initiative Program—NCN17_129.

Institutional Review Board Statement: The study was conducted according to the guidelines of the Declaration of Helsinki and was approved by the Ethics Committee at Pontificia Universidad Catolica (protocol code 171127001, 20 June 2019).

Informed Consent Statement: Informed consent was not obtained for this study as the data were used retrospectively and obtained in a previous study.

Data Availability Statement: Data are available upon request to the corresponding author.

Conflicts of Interest: The authors declare no conflict of interest.

Appendix A

Bland–Altman Plots

Table A1. The Bland–Altman plots for the mean and standard deviation (Bland–Altman plots show 1.96 × SD).

Hemodynamic Parameter	Section I Mean (SD)	Section II Mean (SD)	Section III Mean (SD)	Section IV Mean (SD)
Velocity (m/s)	0.034 (0.080)	0.045 (0.058)	0.057 (0.064)	0.083 (0.050)
WSS (N/m^2)	0.055 (0.182)	0.074 (0.136)	0.132 (0.132)	0.197 (0.112)
WSSA (N/m^2)	0.056 (0.179)	0.073 (0.135)	0.132 (0.132)	0.197 (0.112)
WSSC (N/m^2)	−0.001 (0.035)	0.007 (0.014)	−0.001 (0.008)	−0.004 (0.014)
Vorticity (1/s)	−0.289 (6.074)	2.279 (8.932)	2.983 (7.033)	7.531 (8.296)
Viscous Dissipation (1/s^2)	−0.257 (1.771)	0.436 (0.705)	0.246 (0.587)	0.591 (0.964)
Kinetic Energy (μJ)	−0.013 (0.102)	0.026 (0.040)	0.015 (0.032)	0.032 (0.052)
Energy Loss (μW)	0.301 (0.870)	0.404 (0.581)	0.472 (0.455)	0.689 (0.498)

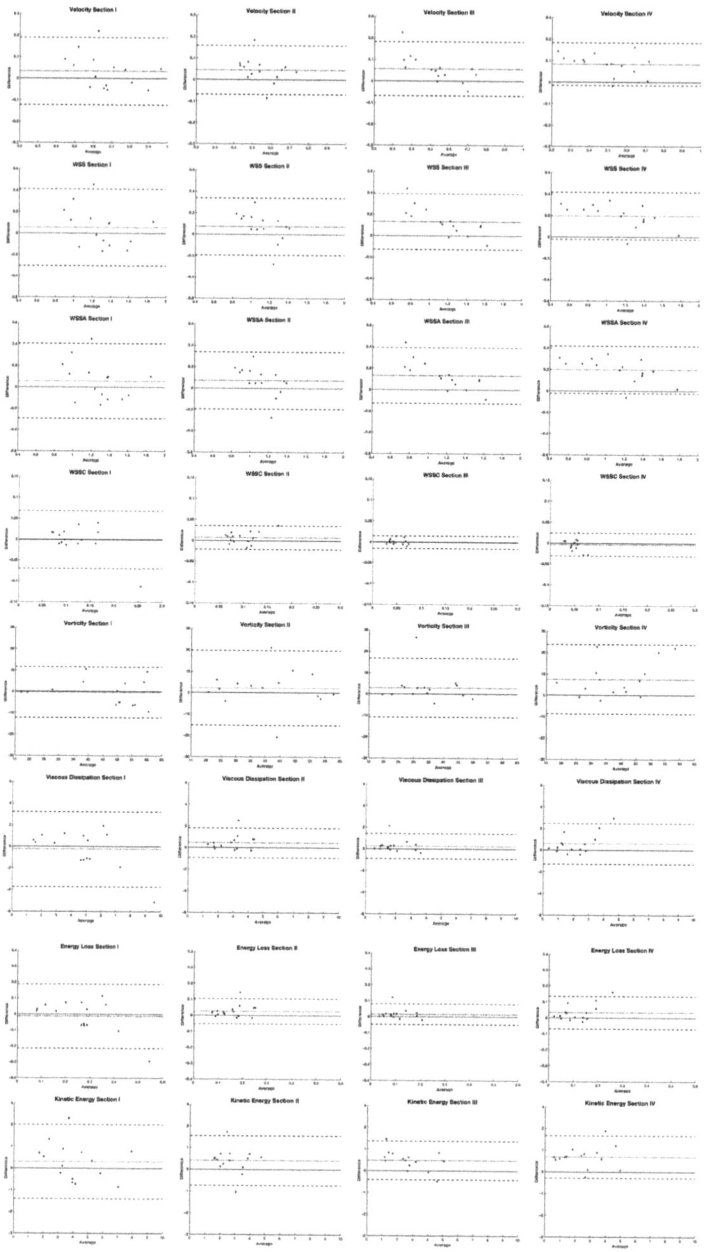

Figure A1. The Bland–Altman plots for comparing the hemodynamic parameters computed in this research, with and without self-gating: velocity; WSS; WSSA; WSSC; vorticity; viscous dissipation; energy loss; and kinetic energy. The average axis of the Bland–Altman plots corresponds to the average of the data with self-gating (SG) and without self-gating (NG), $Average = \frac{SG+NG}{2}$. The difference axis corresponds to the difference between the data with self-gating (SG) and without self-gating (NG), $Difference = SG - NG$. The difference axis shows the lines plotting the mean $\pm$ 1.96 times the standard deviation, and the zero value for comparison purposes.

Appendix B

Local Differences

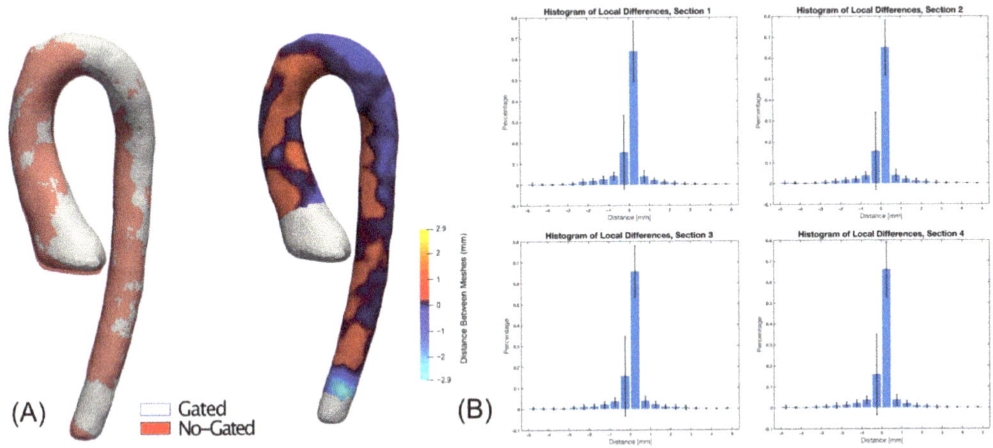

Figure A2. (**A**) The local differences of the respiratory self-gated and non-gated segmentations in a representative volunteer. The differences were calculated by measuring the Euclidean distance between each triangle of the self-gated surface mesh to the nearest triangle of the non-gated mesh along the normal direction of the first one. The region where the aorta was segmented from the left ventricle showed more significant differences between the segmentations, with and without self-gating. (**B**) A histogram of the local differences calculated at each section for all volunteers. The bar plots show the fraction of differences calculated in each section and classified by its distance in millimeters. The local differences obtained between ± 1.0 mm were 87.61% in section I, 88.27% in section II, 88.66% in section III, and 89.11% in section IV.

Table A2. The mean values of the local differences between the respiratory self-gated and non-gated segmentations, computed at each section using the signed value differences.

Hemodynamic Parameter	Section I	Section II	Section III	Section IV
Mean (mm)	−0.0811	−0.0706	−0.0505	−0.0716
Standard deviation (mm)	0.7883	0.7685	0.7548	0.7399
Minimum value (mm)	−5.1363	−4.7774	−4.7275	−4.6676
Maximum value (mm)	2.9155	2.6580	2.6624	2.9427

References

1. Dyverfeldt, P.; Ebbers, T. Comparison of respiratory motion suppression techniques for 4D flow MRI. *Magn. Reson. Med.* **2017**, *78*, 1877–1882. [CrossRef]
2. Stehning, C.; Börnert, P.; Nehrke, K.; Eggers, H.; Stuber, M. Free-breathing whole-heart coronary MRA with 3D radial SSFP and self-navigated image reconstruction. *Magn. Reson. Med.* **2005**, *54*, 476–480. [CrossRef] [PubMed]
3. Liao, J.-R.; Pauly, J.M.; Brosnan, T.J.; Pelc, N.J. Reduction of motion artifacts in cine MRI using variable-density spiral trajectories. *Magn. Reson. Med.* **1997**, *37*, 569–575. [CrossRef] [PubMed]
4. Uribe, S.; Beerbaum, P.; Sørensen, T.S.; Rasmusson, A.; Razavi, R.; Schaeffter, T. Four-dimensional (4D) flow of the whole heart and great vessels using real-time respiratory self-gating. *Magn. Reson. Med.* **2009**, *62*, 984–992. [CrossRef]
5. Henningsson, M.; Koken, P.; Stehning, C.; Razavi, R.; Prieto, C.; Botnar, R.M. Whole-heart coronary MR angiography with 2D self-navigated image reconstruction. *Magn. Reson. Med.* **2012**, *67*, 437–445. [CrossRef]
6. Weiger, M.; Börnert, P.; Proksa, R.; Schäffter, T.; Haase, A. Motion-adapted gating based onk-space weighting for reduction of respiratory motion artifacts. *Magn. Reson. Med.* **1997**, *38*, 322–333. [CrossRef] [PubMed]
7. Scott, A.D.; Keegan, J.; Firmin, D.N. High-resolution 3D coronary vessel wall imaging with near 100% respiratory efficiency using epicardial fat tracking: Reproducibility and comparison with standard methods. *J. Magn. Reson. Imaging* **2011**, *33*, 77–86. [CrossRef] [PubMed]

8. Akcakaya, M.; Gulaka, P.; Basha, T.A.; Ngo, L.H.; Manning, W.J.; Nezafat, R. Free-breathing phase contrast MRI with near 100% respiratory navigator efficiency using k-space-dependent respiratory gating. *Magn. Reson. Med.* **2014**, *71*, 2172–2179. [CrossRef] [PubMed]
9. Makowski, M.R.; Wiethoff, A.J.; Jansen, C.H.P.; Uribe, S.; Parish, V.; Schuster, A.; Botnar, R.M.; Bell, A.; Kiesewetter, C.; Razavi, R.; et al. Single breath-hold assessment of cardiac function using an accelerated 3D single breath-hold acquisition technique—Comparison of an intravascular and extravascular contrast agent. *J. Cardiovasc. Magn. Reson.* **2012**, *14*, 53. [CrossRef] [PubMed]
10. Markl, M.; Frydrychowicz, A.; Kozerke, S.; Hope, M.; Wieben, O. 4D flow MRI. *J. Magn. Reson. Imaging* **2012**, *36*, 1015–1036. [CrossRef] [PubMed]
11. Bastkowski, R.; Weiss, K.; Maintz, D.; Giese, D. Self-gated golden-angle spiral 4D flow MRI. *Magn. Reson. Med.* **2018**, *80*, 904–913. [CrossRef] [PubMed]
12. Nordmeyer, S.; Riesenkampff, E.; Crelier, G.; Khasheei, A.; Schnackenburg, B.; Berger, F.; Kuehne, T. Flow-sensitive four-dimensional cine magnetic resonance imaging for offline blood flow quantification in multiple vessels: A validation study. *J. Magn. Reson. Imaging* **2010**, *32*, 677–683. [CrossRef] [PubMed]
13. Markl, M.; Harloff, A.; Bley, T.A.; Zaitsev, M.; Jung, B.; Weigang, E.; Langer, M.; Hennig, J.; Frydrychowicz, A. Time-resolved 3D MR velocity mapping at 3T: Improved navigator-gated assessment of vascular anatomy and blood flow. *J. Magn. Reson. Imaging* **2007**, *25*, 824–831. [CrossRef] [PubMed]
14. van Ooij, P.; Semaan, E.; Schnell, S.; Giri, S.; Stankovic, Z.; Carr, J.; Barker, A.J.; Markl, M. Improved respiratory navigator gating for thoracic 4D flow MRI. *Magn. Reson. Imaging* **2015**, *33*, 992–999. [CrossRef]
15. Markl, M.; Chan, F.P.; Alley, M.T.; Wedding, K.L.; Draney, M.T.; Elkins, C.J.; Parker, D.W.; Wicker, R.; Taylor, C.A.; Herfkens, R.J.; et al. Time-resolved three-dimensional phase-contrast MRI. *J. Magn. Reson. Imaging* **2003**, *17*, 499–506. [CrossRef]
16. Montalba, C.; Urbina, J.; Sotelo, J.; Andia, M.E.; Tejos, C.; Irarrazaval, P.; Hurtado, D.E.; Valverde, I.; Uribe, S. Variability of 4D flow parameters when subjected to changes in MRI acquisition parameters using a realistic thoracic aortic phantom. *Magn. Reson. Med.* **2018**, *79*, 1882–1892. [CrossRef]
17. Sotelo, J.; Mura, J.; Hurtado, D.; Uribe, S. A Novel MATLAB Toolbox for Processing 4D Flow MRI Data. *Int. Soc. Magn. Reson. Med.* **2019**, *27*, 2.
18. Sotelo, J.; Urbina, J.; Valverde, I.; Tejos, C.; Irarrazaval, P.; Andia, M.E.; Uribe, S.; Hurtado, D.E. 3D Quantification of Wall Shear Stress and Oscillatory Shear Index Using a Finite-Element Method in 3D CINE PC-MRI Data of the Thoracic Aorta. *IEEE Trans. Med. Imaging* **2016**, *35*, 1475–1487. [CrossRef]
19. Sotelo, J.; Dux-Santoy, L.; Guala, A.; Rodríguez-Palomares, J.; Evangelista, A.; Sing-Long, C.; Urbina, J.; Mura, J.; Hurtado, D.E.; Uribe, S. 3D axial and circumferential wall shear stress from 4D flow MRI data using a finite element method and a laplacian approach. *Magn. Reson. Med.* **2018**, *79*, 2816–2823. [CrossRef]
20. Sotelo, J.; Urbina, J.; Valverde, I.; Mura, J.; Tejos, C.; Irarrazaval, P.; Andia, M.E.; Hurtado, D.E.; Uribe, S. Three-dimensional quantification of vorticity and helicity from 3D cine PC-MRI using finite-element interpolations. *Magn. Reson. Med.* **2018**, *79*, 541–553. [CrossRef] [PubMed]
21. Sotelo, J.; Valverde, I.; Martins, D.; Bonnet, D.; Boddaert, N.; Pushparajan, K.; Uribe, S.; Raimondi, F. Impact of aortic arch curvature in flow haemodynamics in patients with transposition of the great arteries after arterial switch operation. *Eur. Heart J. Cardiovasc. Imaging* **2021**, jeaa416. [CrossRef] [PubMed]
22. Arrieta, C.; Uribe, S.; Ramos-Grez, J.; Vargas, A.; Irarrazaval, P.; Parot, V.; Tejos, C. Quantitative assessments of geometric errors for rapid prototyping in medical applications. *Rapid Prototyp. J.* **2012**, *18*, 431–442. [CrossRef]

applied sciences

Article

Comprehensive Assessment of Left Intraventricular Hemodynamics Using a Finite Element Method: An Application to Dilated Cardiomyopathy Patients

Pamela Franco [1,2,3], Julio Sotelo [1,3,4], Cristian Montalba [1,3,5], Bram Ruijsink [6,7], Eric Kerfoot [6], David Nordsletten [8,9], Joaquín Mura [3,10], Daniel Hurtado [3,11,12] and Sergio Uribe [1,3,5,*]

1 Biomedical Imaging Center, School of Engineering, Pontificia Universidad Católica de Chile, Santiago 7820436, Chile; pafranco@uc.cl (P.F.); julio.sotelo@uv.cl (J.S.); cristian.montalba@uc.cl (C.M.)
2 Electrical Engineering Department, School of Engineering, Pontificia Universidad Católica de Chile, Santiago 7820436, Chile
3 Millennium Nucleus for Cardiovascular Magnetic Resonance, Santiago 7820436, Chile; joaquin.mura@usm.cl (J.M.); dhurtado@ing.puc.cl (D.H.)
4 School of Biomedical Engineering, Universidad de Valparaíso, Valparaíso 2362905, Chile
5 Radiology Department, School of Medicine, Pontificia Universidad Católica de Chile, Santiago 833023, Chile
6 Guy's and St Thomas' NHS Foundation Trust, London SE1 7EH, UK; jacobus.ruijsink@kcl.ac.uk (B.R.); eric.kerfoot@kcl.ac.uk (E.K.)
7 School of Imaging Sciences and Biomedical Engineering, Faculty of Science of Life and Medicine, King's College London, London WC2R 2LS, UK
8 Division of Biomedical Engineering and Imaging Sciences, Department of Biomedical Engineering, King's College London, London WC2R 2LS, UK; nordslett@umich.edu
9 Department of Biomedical Engineering and Cardiac Surgery, University of Michigan, North Campus Research Center, Ann Arbor, MI 48109, USA
10 Department of Mechanical Engineering, Universidad Técnica Federico Santa María, Valparaíso 2390123, Chile
11 Structural and Geotechnical Engineering Department, School of Engineering, Pontificia Universidad Católica de Chile, Santiago 7820436, Chile
12 Institute for Biological and Medical Engineering, School of Engineering, Medicine and Biological Sciences, Pontificia Universidad Católica de Chile, Santiago 7820436, Chile
* Correspondence: suribe@uc.cl

Citation: Franco, P.; Sotelo, J.; Montalba, C.; Ruijsink, B.; Kerfoot, E.; Nordsletten, D.; Mura, J.; Hurtado, D.; Uribe, S. Comprehensive Assessment of Left Intraventricular Hemodynamics Using a Finite Element Method: An Application to Dilated Cardiomyopathy Patients. *Appl. Sci.* **2021**, *11*, 11165. https://doi.org/10.3390/app112311165

Academic Editor: Julio Garcia Flores

Received: 6 October 2021
Accepted: 8 November 2021
Published: 25 November 2021

Abstract: In this paper, we applied a method for quantifying several left intraventricular hemodynamic parameters from 4D Flow data and its application in a proof-of-concept study in dilated cardiomyopathy (DCM) patients. In total, 12 healthy volunteers and 13 DCM patients under treatment underwent short-axis cine b-SSFP and 4D Flow MRI. Following 3D segmentation of the left ventricular (LV) cavity and registration of both sequences, several hemodynamic parameters were calculated at peak systole, e-wave, and end-diastole using a finite element approach. Sensitivity, inter- and intra-observer reproducibility of hemodynamic parameters were evaluated by analyzing LV segmentation. A local analysis was performed by dividing the LV cavity into 16 regions. We found significant differences between volunteers and patients in velocity, vorticity, viscous dissipation, energy loss, and kinetic energy at peak systole and e-wave. Furthermore, although five patients showed a recovered ejection fraction after treatment, their hemodynamic parameters remained low. We obtained several hemodynamic parameters with high inter- and intra-observer reproducibility. The sensitivity study revealed that hemodynamic parameters showed a higher accuracy when the segmentation underestimates the LV volumes. Our approach was able to identify abnormal flow patterns in DCM patients compared to volunteers and can be applied to any other cardiovascular diseases.

Keywords: 4D flow MRI; flow quantification; finite elements; left ventricle; dilated cardiomyopathy

1. Introduction

Dilated cardiomyopathy (DCM) is more common than non-ischemic cardiomyopathy and leads to left ventricular dilation and systolic and diastolic dysfunction [1,2]. The process that alters the heart's size, geometry, and function is associated with increased hemodynamic demands, which cause abnormal mechanical stress in the muscle [3,4]. The progression is associated with an incremented risk of heart failure and sudden cardiac death [5]. However, poor survival and high mortality rate reveal that effective treatment of DCM-related heart failure remains challenging. Pharmacological and resynchronization therapies have improved DCM treatment by halting disease progression and leading to reverse remodeling [1].

The preferred imaging technique for assessing the heat in DCM patients is cardiovascular magnetic resonance (CMR). CMR allows the acquisition of anatomical, cine, and velocity images, including 4D Flow MR [5–8].

4D Flow allows a qualitative and quantitative analysis of several hemodynamic parameters. It has been applied extensively in the great vessels, particularly in the aorta [9–11] and in the left ventricle (LV) for assessing intraventricular flow in some cardiovascular diseases [12–19]. Previous studies have demonstrated that lower kinetic energy values in diastole are associated with the deterioration of ventricular filling, induced by morphological alteration commonly found in Fontan patients, mitral regurgitation, and LV dysfunction or remodeling [13–15]. Additionally, turbulent kinetic energy has shown a stronger association with the ventricle's remodeling in patients with Tetralogy of Fallot and higher values in DCM patients compared with normal subjects [16]. Vortex formation has been studied qualitatively (vortex size and location) and quantitatively (Lagrangian Coherent Structures and the curl of velocity) [17–20]. These studies suggest that parameters associated with the 3D intraventricular flow may be critical for LV filling and ejection and could be relevant to the development of dilation, dysfunction, and prognosis in patients with heart diseases. While these measures have a potential role in describing intraventricular flow, the difficulties of implementing them have led to the analysis of only a few combinations of these parameters in a single cohort of patients [13–15,17,19].

Due to the multidirectional velocity data, although impressively comprehensive, it may need to be supplemented by more selective flow imaging at high temporal and spatial resolutions or computational fluid dynamics simulation. Reaching conclusions regarding small-scale methodology, which comprehensively describes the characteristics of intraventricular flow, could improve the use of intraventricular 4D Flow for clinical research and potential translation to clinical settings.

In this work, we adapted a method for quantifying 4D Flow in the aorta [9–11,20]. We modified the methodology applied in the left ventricle to obtain several hemodynamic parameters from a single segmentation from a 4D Flow dataset and cine MRI. To show the applicability of this approach, we performed a proof-of-concept study in which we applied the method in a small cohort of DCM patients to find which parameters were different from volunteers. We obtained three-dimensional hemodynamic parameters, including kinetic energy, vorticity, helicity density, viscous dissipation, and energy loss [9,13,21–25].

2. Materials and Methods

2.1. Population

A total of 12 healthy volunteers (HV), mean age 40.8 years (range 27–55 years), and 13 DCM patients, mean age 48.7 years (range 29–62 years), matched according to age and gender, were included in this research. Demographical and clinical data are described in Table 1. At the time of diagnosis, DCM was defined as the presence of symptoms and signs of heart failure with echocardiographic signs of ventricular enlargement and systolic myocardial dysfunction in the absence of hypertension, valve diseases, or significant coronary artery diseases sufficient to cause global systolic impairment, by the definition of the European Society of Cardiology [26]. Our DCM cohort all received treatment with an improved LV ejection fraction (range 51–66%) and LV volume indices at CMR imaging. All patients received standard guideline-directed treatment for DCM following the 2008 heart failure guidelines from the European society of cardiology. The details of treatments were not available, as our center is the referral center for several clinics for cardiac CMR. The

HV had normal electrocardiograms and echocardiographic examinations without valvular or ventricular dysfunction. All subjects participated under informed consent, with data collection approved by the Regional Ethics Committee, South East London, UK (REC, 12/LO/1456).

Table 1. Demographical and clinical data for healthy and DCM patients. All quantitative data are expressed as the median (range). HR: Heart Rate, EF: Ejection Fraction, LVSV: Left Ventricle Stroke Volume, CO: Cardiac Output, LVEDV: Left Ventricle End-Diastolic Volume, and Left Ventricle End-Systolic Volume. * indicates statistically significant differences ($p < 0.05$).

				DCM Group		
	HV	**DCM**	***p*-Value**	**LVEF ≥ 50 (Complete-Responders)**	**LVEF < 50 (Non-Responders)**	***p*-Value**
N	12	13		5	8	
Age (years)	39 (27,55)	51 (29,62)	0.060	40 (29,62)	53 (44,58)	0.502
Gender (female:male)	5:7	6:7	0.821	3:2	3:5	0.429
Weight (kg)	68 (50,111)	83 (43,116)	0.213	90 (72,116)	72.5 (43,95)	0.071
Height (cm)	173 (163,188)	168 (155,178)	0.203	168 (163,178)	166.5 (155,175)	0.454
HR (bpm)	64 (58,78)	65 (56,101)	0.743	65 (56,101)	67.5 (57,89)	0.698
EF (%)	62.7 (54,69)	46 (29,66)	<0.001 *	55 (51,66)	44 (29,48)	0.002 *
LVSV (mL)	95.5 (66.3,122.9)	62 (53,132.1)	0.039 *	61 (53,89)	79 (55,132,1)	0.183
CO (L/min)	6.4 (4.8,7.9)	6.1 (4.4,7.9)	0.327	6.3 (5.2,7.9)	5.9 (4.4,7.7)	0.524
LVEDV (mL)	153 (105.6,197.1)	199 (125,364.2)	0.015 *	187 (151,201)	219.5 (125,364.2)	0.050 *
LVESV (mL)	51 (39,88)	92 (37,232.1)	0.004 *	75 (37,92)	125 (68,232.1)	0.045 *

2.2. Data Acquisition

Multi-slice 2D cine balanced steady-state free precession (b-SSFP) and 4D Flow MRI data were acquired in all subjects using a clinical 1.5 T MT Scanner (Philips Achieva, Philips Medical Systems, Best, The Netherlands). During the MRI examination, multi-slice b-SSFP was used to acquire short-axis morphological images in 40 frames with 8 mm slice thickness, using retrospective cardiac gating. Acquisition parameters were echo time (TE) 1.4 ms, repetition time (TR) of 2.8 ms, flip angle (FA) of 60° and acquired and reconstructed pixel sizes were 2.47×2.53 mm^3 and 1.45 mm^2, respectively. 4D Flow MRI data were acquired during free-breathing with MR parameters, as follows: TW of 2.3 ms, TR of 4.7 ms, FA of 6°, velocity encoding of 130 cm/s, and spatial resolution (acquired and reconstructed) $2.5 \times 2.5 \times 2.5$ mm^3. These settings gave a temporal resolution of 58 ms. After the acquisition, the 4D Flow MRI data were reconstructed into 24 cardiac phases on the MRI system.

2.3. Data Analysis

The 4D Flow MRI datasets were processed using an-house MATLAB library (The MathWorks Inc., Natick, MA, USA), which included the registration of the b-SSFP cine and 4D Flow MR images, interpolation of the b-SSFP images, segmentation of the LV, and generation of the finite element mesh (Figure 1).

The Eidolon software was used to perform the registration between the multi-slice b-SSFP and the 4D Flow MRI (King's College London, London, UK) [27]. To obtain a smooth tetrahedral mesh, we doubled the number of slices in the b-SSFP images by using a cubic interpolation of values at neighboring grid points in each respective dimension, obtaining a final voxel size of $1.43 \times 1.43 \times 4.04$ mm^3. LV endocardium was automatically segmented throughout all cardiac phases in the short-axis cine b-SSFP images, using the image analysis software Segment v2.2R6410 (Medviso AB, Lund, Sweden) [28–30]. The segmentation was visually inspected and manually corrected if needed. Segmentations c were then used to generate a binary mask. Afterward, we created a tetrahedral mesh using the iso2mesh MATLAB Toolbox [31]. Once the mesh was constructed, we computed the velocity vector at each mesh node from the 4D Flow datasets using a cubic interpolation. 3D maps of vorticity, helicity density, viscous dissipation, energy loss, and kinetic energy fields were then calculated using a previously published finite element approach [9–11,25]. The description of the equations used to calculate each hemodynamic parameter is presented in Table S1. The parameters were averaged at peak systole, e-wave, and end-diastole using one timeframe before and after to reduce noise in the data.

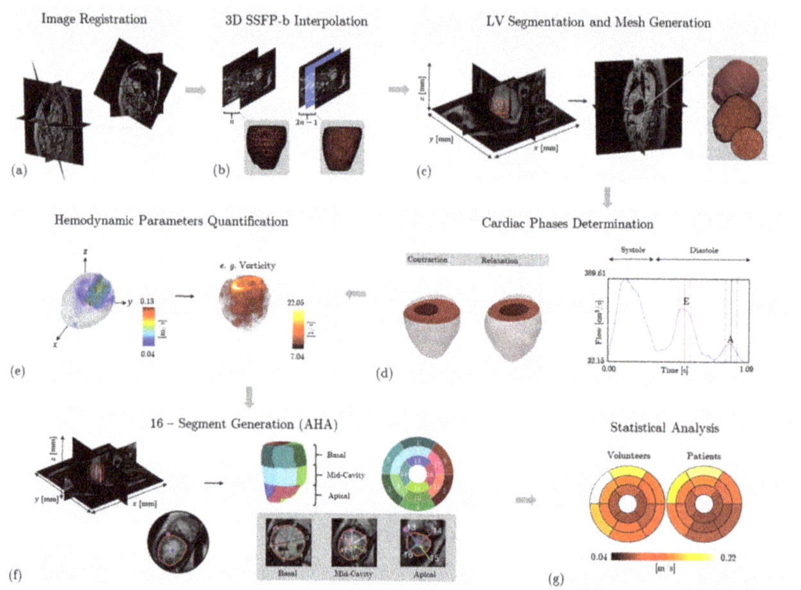

Figure 1. Schematic description of the quantification process. (**a**) First, we performed registration of the 4D Flow with the b-SSFP images. (**b**) Second, we doubled the number of slices in the b-SSFP images, (**c**) then the LV segmentation and tetrahedral mesh were generated. (**d**) We estimated the cardiac phases under study (**e**) and we transferred the velocity information at each node of the mesh from the 4D Flow MRI datasets using cubic interpolation. (**f**) Then, we calculated hemodynamic parameters under study. (**g**) Finally, the mean values of hemodynamic parameters were included in a bullseye plot to compare volunteers and DCM patients.

2.4. Local Hemodynamics

A 16-segment model was used to divide the LV. In contrast to a standardized nomenclature, a minor adjustment was made [26]. Due to the generally intricate shape of the apical region of the LV, region 17 was excluded from our analysis. Accordingly, LV mesh was divided into 16 segmentations. The centerline of the LV was calculated automatically by detecting the centroid of the LV contour in each slice and connected to create a line. To determine the three sections of the LV, we divided the centerline into three equal parts perpendicular to the long axis of the heart. An additional point was then manually placed at the junction between the right ventricular free wall and the interventricular septum on the LV. Based on these positions, landmarks were uniformly distributed along the boundaries. Each section was then partitioned into six segments of 60° each on basal and mid-cavity sections and four segments of 90° each on apical section (Figure 1f). Finally, for visualization purposes, we used the scientific software ParaView version 5.3.0 (Kitware, Clifton Park, NY, USA).

2.5. Statistical Analysses

Normal distribution in population demographics was evaluated using the Shapiro-Wilk test. Differences between groups for continuous parameters were assessed by a Student *t*-test if they presented a normal distribution, and the Mann-Whitney *U* test otherwise. The χ^2 test was applied for categorical variables, which were reported as percentages. A *p*-value < 0.05 was considered statistically significant. The statistical analyses were performed using GraphPad Prism version 6.0.1 (GraphPad Software Inc., San Diego, CA, USA).

These data were displayed in box-whisker and bullseye plots for global and local analyses, respectively. Additionality, a correlation matrix-based hierarchical clustering method was introduced to extract multiple correlation patterns from hemodynamic parameters. This method can effectively identify highly correlated data. The results are described with a tree structure plot called a dendrogram. The present study used Pearson's correlation method to measure the similarity between hemodynamic parameters [32].

Furthermore, a sensitivity study was performed by looking at changes in the hemodynamic parameters subjected to the LV segmentation changes. We increased and decreased the size of the LV cavity from the first segmentations by moving the segmentation contour in 0.5 to 2 pixels of the b-SSFP image, equivalent to 0.72 to 2.89 mm. We compared the results with the original LV segmentation's respective mean value at each cardiac phase studied. We used the Kruskal-Wallis test to compare the variables across the different LV segmentation, with a p-value < 0.05 indicating statistical significance. The significance level was adjusted by using Dunn's test correction.

To assess the inter-observer agreement, data were analyzed by two independent observers, one with three years of experience in MR LV quantification and the other a medical technologist with no previous experience in this field. In addition, re-analyzed images with a 1-month interval to evaluate the intra-observer reproducibility. Inter- and intra-observer reproducibility were analyzed using Bland-Altman plots, and the results are shown in the Supplementary Materials.

3. Results

3.1. Study Population

There were no significant differences in age and heart rate (Table 1). However, ejection fraction and stroke volume were lower in DCM patients than volunteers, while end-diastolic and end-systolic volumes were larger. These changes indicated that the LV in DCM patients was enlarged and its cardiac function was reduced, which is consistent with the pathological characteristics of DCM [1,2]. Additionally, eight patients still showed significantly impaired systolic function at CMR's time (non-responders), and five patients showed a complete response to treatment (complete responders). Between DCM groups, complete- vs. non-responders, there were differences in ejection fraction, end-diastolic volume, and end-systolic volume.

3.2. Global Hemodynamics

Assessment of global hemodynamic parameters is shown in Figure 2 and Table 2. Volunteers showed higher hemodynamics values than patients at peak systole and e-wave, except for helicity density. Remarkably, hemodynamic parameters in complete responder DCM patients remained low compared to volunteers. We found statistical differences between HV and DCM patients: non- and complete responders at peak systole and e-wave in velocity, vorticity, viscous dissipation, energy loss, and kinetic energy. In all cases, p-values were lower or equal to 0.005. There were no statistical differences in the parameters at end-diastole. Furthermore, we did not find statistical differences between DCM groups. In addition, ROC curves showed that previous parameters discriminated between HV and DCM patients (Figure S1).

The total computational time used to process the data, once the multi-slice b-SSFP was segmented and registered, varied between 30–40 s for one cardiac phase, using a standard computer (3.4 GHz Intel ® Core i7™, 16 GB RAM).

Hierarchical cluster analysis (Figure 3) provides an alternative method for reliable identification of correlation between ejection fraction and hemodynamics parameters from 4D-flow MRI. According to their similarities, they are classified into two clusters identified at peak-systole, e-wave, and end-diastole. At peak-systole and e-wave, cluster 1 (black): helicity density; and cluster 2 (red): ejection fraction, energy loss, vorticity, viscous dissipation, velocity, and kinetic energy. Finally, at end-diastole, cluster 1 (red): helicity density and ejection fraction, and cluster 2 (red): energy loss, viscous dissipation, vorticity, velocity, and kinetic energy. This means that ejection fraction correlates with all parameters except helicity density at peak systole and e-wave.

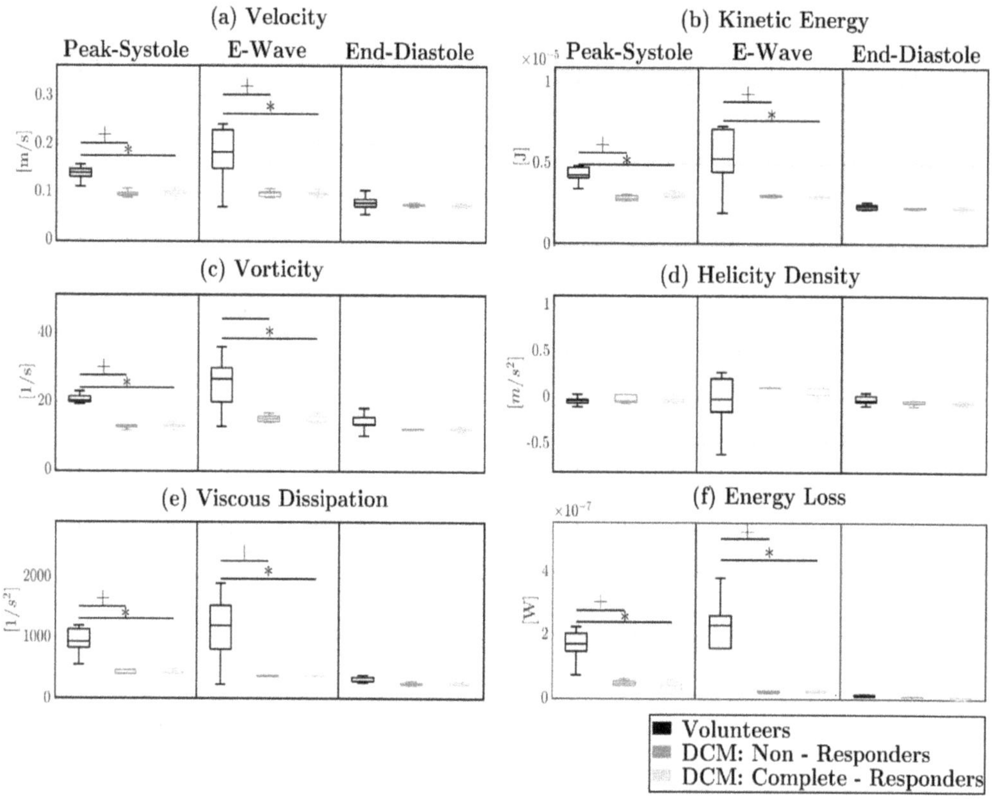

Figure 2. Box whisker plots for hemodynamic parameters (**a–f**) in the entire LV cavity of HV and DCM patients groups at peak systole, e-wave, and end-diastole. On each box, the central mark is the median, the bottom and top edges of the box are the 25th and 75th percentiles, respectively, and the whiskers extend to the most extreme data points not considered outliers. *,+ Indicates statistically significant differences ($p < 0.05$).

3.3. Sensitivity Study, Intra-, and Inter-Observer Reproducibility

Figure 4 shows a sensitivity analysis at peak-systole. The relative error in LV cardiac volumes and helicity density did not show significant differences between groups across LV segmentation. There were significant differences for some segmentation for the other parameters, particularly for the velocity magnitude, energy loss, and kinetic energy. The hemodynamic parameters showed a relative error proportional to the dilatation or erosion of the contour in the segmentation. When the segmentation was dilated or eroded 1 pixel or less, the relative error differences, with respect to the original segmentation for volunteers and DCM patients, were: velocity magnitude (9.03%, 6.78%), vorticity magnitude (5.49%, 2.74%), helicity density (12.11%, 13.98%), viscous dissipation (6.31%, 3.34%), energy loss (3.59%, 5.89%), and kinetic energy (7.66%, 6.09%). Similar results were obtained at e-wave and end-diastole, as shown in Figures S2 and S3. Those errors were more significant, particularly when the segmentation was dilated or eroded by more than 1 pixel. Helicity density and energy loss showed greater dependency on the segmentation error.

Regarding reproducibility, there was an excellent agreement of inter- and intra-observer analysis of global hemodynamic parameters. Details are given in the Appendix A.

Table 2. Global hemodynamics data for HV, complete, and non-responders DCM patients. All quantitative data are expressed as the mean ± standard deviation. *,+ Indicates statistically significant differences ($p < 0.05$).

	HV	DCM Group		p-Value	
		Complete-Responders	Non-Responders	HV vs. Complete-Responders	HV vs. Non-Responders
Peak-systole					
Velocity (m/s)	0.140 ± 0.014	0.099 ± 0.007	0.096 ± 0.008	<0.001 *	<0.001 +
Kinetic Energy (μJ)	43.722 ± 4.592	29.335 ± 1.917	31.288 ± 2.044	<0.001 *	<0.001 +
Vorticity (1/s)	20.306 ± 2.075	12.934 ± 0.814	13.331 ± 1.251	<0.001 *	<0.001 +
Helicity Density (m/s²)	−0.042 ± 0.004	0.036 ± 0.161	−0.077 ± 0.139	0.125	0.417
Viscous Dissipation (1/s²)	970.840 ± 412.093	412.093 ± 61.107	421.080 ± 54.870	<0.001 *	<0.001 +
Energy Loss (ηW)	173.080 ± 39.387	35.284 ± 14.144	47.734 ± 12.935	<0.001 *	<0.001 +
E-wave					
Velocity (m/s)	0.187 ± 0.059	0.097 ± 0.004	0.099 ± 0.014	0.007 *	<0.001 +
Kinetic Energy (μJ)	5.567 ± 1.810	3.008 ± 0.074	3.005 ± 0.423	0.007 *	<0.001 +
Vorticity (1/s)	26.309 ± 7.895	14.633 ± 0.755	14.931 ± 1.963	0.005 *	<0.001 +
Helicity Density (m/s²)	0.106 ± 0.441	0.077 ± 0.065	0.056 ± 0.119	0.907	0.785
Viscous Dissipation (1/s²)	1208.091 ± 574.696	393.994 ± 26.632	370.314 ± 80.785	0.007 *	<0.001 +
Energy Loss (ηW)	217.440 ± 126.751	28.408 ± 6.340	24.329 ± 11.540	0.005 *	<0.001 +
End-Diastole					
Velocity (m/s)	0.075 ± 0.015	0.077 ± 0.005	0.073 ± 0.002	0.796	0.673
Kinetic Energy (μJ)	2.291 ± 0.423	2.359 ± 0.165	2.266 ± 0.063	0.739	0.871
Vorticity (1/s)	13.629 ± 2.255	12.444 ± 0.633	12.203 ± 0.552	0.273	0.099
Helicity Density (m/s²)	−0.025 ± 0.084	−0.082 ± 0.064	−0.027 ± 0.096	0.201	0.979
Viscous Dissipation (1/s²)	300.577 ± 89.250	289.131 ± 54.300	256.496 ± 16.779	0.795	0.188
Energy Loss (ηW)	11.162 ± 7.191	10.459 ± 7.757	8.875 ± 1.896	0.859	0.395

3.4. Local Hemodynamics

Figure 5 shows the bullseye plots of the hemodynamic parameters for volunteers and DCM patients at peak systole. More areas with statistical differences were observed mainly in velocity magnitude and kinetic energy, particularly in anteroseptal, inferior, inferolateral basal, anterior, inferoseptal, inferior, inferolateral mid-cavity, anterior, septal, and lateral apical segments (all p-values < 0.033). Additionally, vorticity magnitude showed statistical differences in anteroseptal basal ($p = 0.045$) and inferoseptal mid-cavity ($p = 0.046$) segments. Energy loss showed statistical differences in inferoseptal ($p = 0.029$) and inferolateral ($p = 0.046$) mid-cavity segments and in anterior ($p = 0.023$), septal ($p = 0.070$), and inferior ($p = 0.077$) apical segments. Helicity density and viscous dissipation did not show statistical differences in any parcellation.

Figure S4 shows the comparison at the e-wave. Similar to peak systole, statistical differences were in velocity magnitude and kinetic energy. Statistical differences in viscous dissipation and energy loss were found in inferolateral basal and septal, anterior, and lateral apical segments (all p-values < 0.049). Vorticity magnitude showed statistical differences in anteroseptal basal ($p = 0.049$) and septal ($p = 0.039$) and lateral ($p = 0.039$) apical segments. Helicity density did not show statistical differences in any parcellation of the LV.

We did not find statistical differences in any segment at end-diastole.

The mean values of local hemodynamic parameters for both groups under study at peak systole, e-wave, and end-diastole are available in Tables S2–S4, respectively.

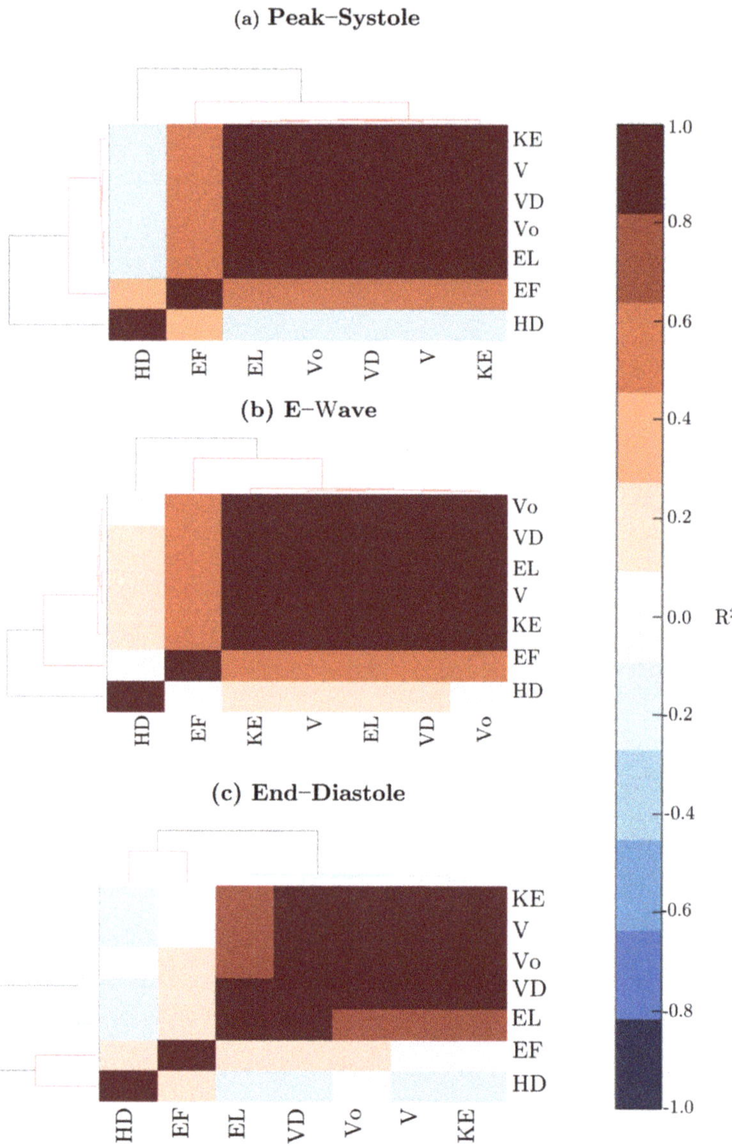

Figure 3. Dendrogram and hierarchical clustering results based on average linkage method for ejection fraction and hemodynamic parameters. EF: ejection fraction, V: velocity, KE: kinetic energy, Vo: vorticity, HD: helicity density, VD: viscous dissipation, and EL: energy loss.

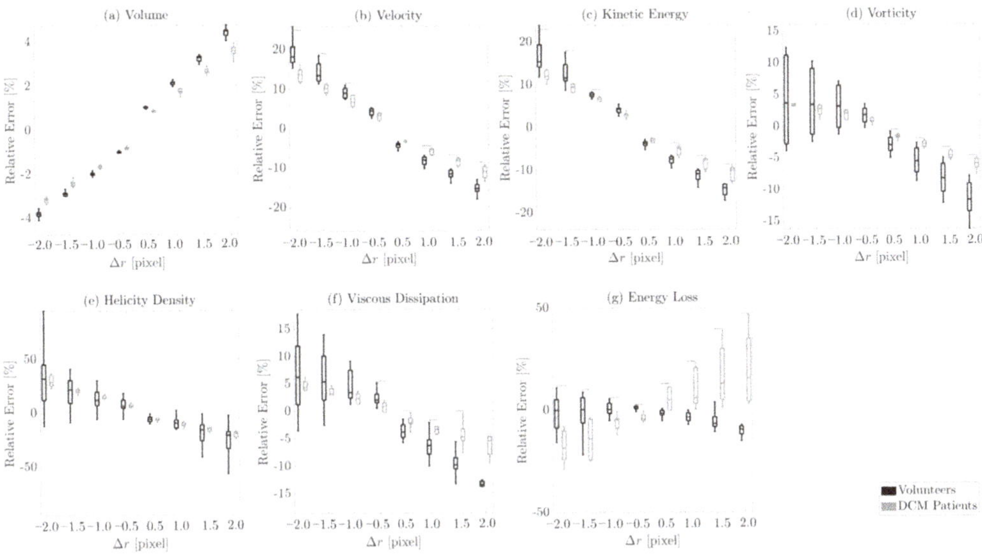

Figure 4. Relative error values of the volume (**a**) and each hemodynamic parameter (**b–g**) obtained comparing the reference segmentation with segmentations given by erosion or dilatation for each group of volunteers and patients at peak systole. * indicates statistically significant differences ($p < 0.05$).

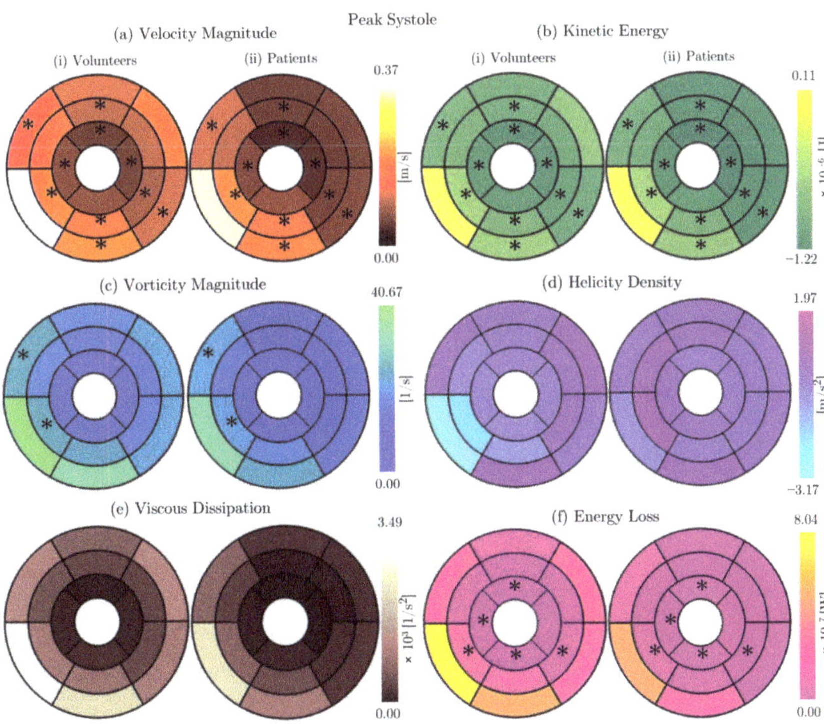

Figure 5. Bullseye plots of mean hemodynamic parameters (**a–f**) across 16 segments for volunteers (i) and patients (ii) at the peak systole. * indicates statistically significant differences ($p < 0.05$).

4. Discussion

We developed a method to characterize the left intraventricular hemodynamics in the LV from 4D Flow MRI using a finite element method, applied in a cohort of DCM patients. This approach estimates vorticity, helicity density, viscous dissipation, energy loss, and kinetic energy fields from a single segmentation. The hemodynamics results indicated that velocity magnitude, vorticity magnitude, viscous dissipation, energy loss, and kinetic energy revealed statistical differences between volunteers and patients, particularly at peak systole and e-wave.

Some of the parameters reported in this study have been reported before. Nevertheless, those parameters have been obtained from different methodologies in different cohorts of patients. In our case, we calculated several parameters from a single segmentation at once from only one 4D flow dataset, which is difficult to determine with other methods.

Some other methods are based on a finite difference approach, as in Lorenz et al. [33]. However, it is well known that finite difference cannot effectively handle complex geometries, such as those found in the cardiovascular system. Neither can impose boundary conditions on irregular surfaces directly but they are both sensitive to noise. Fouras et al. showed that this approach suffers from a loss of accuracy in estimating hemodynamic parameters due to the omission of out-of-plane velocity information [34]. On the other hand, Sotelo et al. demonstrated the convergence and robustness of the finite element method in cardiovascular flow [9,10]. Further, they also showed that the finite element method is both stable and accurate in the presence of noise.

Although DCM mainly affects the systolic function, we evaluated the hemodynamic parameters at systole and diastole, as several papers have shown that diastolic function is also affected by this disease. For instance, Friedberg et al. and Dragulescu et al. reported that diastolic wall-motion abnormalities are prevalent in pediatric DCM. Their presence is associated with diastolic ventricular dysfunction and adverse outcomes [35,36]. Some papers have assessed diastolic function in DCM patients using 4D Flow data [37,38]. They have described alterations in the flow components related to velocity, vorticity, and kinetic energy in a different cohort of patients, consistent with our results [21–25].

As we showed in this study, intraventricular flow in DCM patients was altered compared to healthy volunteers at diastole. In addition, it is interesting to observe that, while end-diastolic volume was significantly larger in patients than healthy subjects, the maximum hemodynamic values for e-wave and end-diastole were smaller in patients than in volunteers. For instance, we found that in the normal LV, kinetic energy values were high. The highest kinetic energy values were observed during early diastole and regionally distributed near basal LV regions. In contrast, early- and end-diastole kinetic energy was lower than normal in a heterogeneous group of DCM patients and decreased with the LV volume. As we found in our results, this decrease of kinetic energy throughout diastole is associated with viscous dissipation and energy loss. That agrees with a previous study, where comparisons of inflow characteristics in healthy subjects and DCM patients showed more differences at e-wave between the two groups [36–41]. These ventricular diastolic function aspects can be influenced by dynamic load and contractility; these may vary within the spectrum of normal conditions. Furthermore, large ventricles lead to weak suction pumps and have high wall tension, which has been previously suggested to cause energy waste and alter vortex ring dynamics [22,23]. These results indicate that, despite the complex nature of ventricular finding factors, clinically useful information regarding left ventricular diastolic function is associated with distinct mitral flow velocity patterns. Therefore, these alterations of blood flow may be a factor in developing systolic and diastolic dysfunctions.

While eight patients showed significantly impaired systolic function at CMR's time, five showed a complete response to treatment. Despite a nearly normalized LV ejection fraction, it showed similar hemodynamic values to non-responders DCM patients that were markedly different from volunteers. These results suggest a significant increase in the ratio of outflow to inflow during systole in responders DCM patients, but the volumes were significantly smaller. Therefore, LV ejection fraction cannot reflect subtle ventricular dysfunction, which potentially can be better assessed using flow-based parameters because of the sensitivity to abnormal pumping function [36]. Therefore, the problem of using LV ejection fraction as the pivotal risk marker for DCM patients is that this single parameter does not recapitulate the complexity of the disease.

The location and extent of the changes in intraventricular blood flow, for example, the depth (base to apex) of the vorticity changes or the spread of impaired flow through the ventricular cavity, can potentially be a sensitive marker of the severity of diseases or the progress of the treatment but are hard to quantify because of the 3D nature of the flow. We proposed to use the bullseye plots to depict this data. These plots allow us to display the most important regional differences and extend flow changes in a familiar way to many clinicians. These results could facilitate homogeneity among 4D Flow quantifiable analysis for clinical researchers and clinicians.

The sensitivity study showed a significant relative error, particularly in helicity density, when the differences in the segmentations were greater than 1 pixel in dilation and erosion cases. We performed this sensitivity study even under the pixel resolution of the 4D Flow data. In each pixel, there were four or five elements from the mesh, whose flow values were interpolated from neighborhood pixels. When the segmentation error was lower or equal to 1 pixel, the maximum mean relative error was less than 10% in most hemodynamic parameters studied. Previous research has shown that DCM patients have a lower mean value of velocity magnitude than healthy volunteers [21–25]. These results were considered when the segmentation contour fell inside the LV blood pool. In general, we also observed that lower errors were obtained for almost all hemodynamic parameters when the segmentation underestimated the LV volumes.

Inter-observer and intra-observer assessments showed excellent reproducibility of the results with negligible mean differences and small limits of an agreement at peak systole, e-wave, and end-diastole for all the parameters assessed. It is important to note that the high intra- e inter-observer variability was obtained because we performed an automatic segmentation process using the software Segment. Therefore, the difference in segmentations was minimal, as previously reported by Tufvesson et al. [29]. The automated process corrections were also minimal, which led to a high intra- and inter-observer variability. On the other hand, the sensitivity study was performed by modifying a reference segmentation in the entire contour by applying erosion or dilatation. This result implies a more significant volume difference concerning the reference, and, as a result, high sensitivity to the segmentation was obtained.

The limited size of this proof-of-concept study did not allow us to investigate the prognostic impact, but this will be our aim in future research. Nevertheless, in this small cohort of patients, we have shown that the velocity, vorticity, kinetic energy, viscous dissipation, and energy loss revealed statistical differences between volunteers and patients. This finding could be relevant to assess changes in a longitudinal study or to study the response to a particular therapy. Additionally, 4D Flow derived parameters showed that, in responding DCM patients, hemodynamics parameters were low, even though they had a recovered ejection fraction. Nevertheless, hierarchical cluster analysis underlined that a moderate correlation may exist between ejection fraction and 4D Flow-based metrics, which needs to be studied further. A clinical study involving more DCM patients should be performed in order to corroborate a prognostic impact—and hence a clinical relevance—of 4D Flow analysis in monitoring DCM patients.

The segmentation of the data need was performed over the multi-slice b-SSFP. Additionally, multi-slice b-SSFP and 4D Flow images need to be registered before analyzing the 4D Flow data. Ideally, the segmentation would be made directly on the 4D Flow data. However, the contrast between the blood pool and myocardium in our 4D Flow data was insufficient to perform accurate segmentation. New sequence developments will likely improve contrast in 4D Flow acquisitions, potentially allowing direct segmentation from the 4D Flow data.

5. Conclusions

This study describes a methodology for quantitative evaluation of intraventricular hemodynamics using a single segmentation from a 4D Flow dataset. We demonstrate that velocity, vorticity, viscous dissipation, energy loss, and kinetic energy can characterize changes in intraventricular flow in DCM patients compared to healthy volunteers. Further studies should focus on the impact of different treatments of DCM patients on those parameters. Our evidence shows that, although ejection fraction may be recovered, the hemodynamic parameters remain low.

Supplementary Materials: The following are available online at https://www.mdpi.com/article/10.3390/app112311165/s1. Table S1: Equations used to calculate each hemodynamic parameter; Table S2: Mean parameter values across 16 segments of the LV, during peak systole, where the bold type is statistically significant between volunteers and patients ($p < 0.05$). υ: Velocity magnitude, ω: vorticity magnitude, H_d: helicity density, VD: viscous dissipation, EL: energy loss, and K: kinetic energy; Table S3: Mean parameter values across 16 segments of the LV during e-wave, where the bold type is statistically significant between volunteers and patients ($p < 0.05$). υ: Velocity magnitude, ω: vorticity magnitude, H_d: helicity density, VD: viscous dissipation, EL: energy loss, and K: kinetic energy; Table S4: Mean parameter values across 16 segments of the LV, during end-diastole, where the bold type is statistically significant between volunteers and patients ($p < 0.05$). υ: Velocity magnitude, ω: vorticity magnitude, H_d: helicity density, VD: viscous dissipation, EL: energy loss, and K: kinetic energy. Figure S1: ROC-curves for hemodynamic parameters (a–f) in the entire LV cavity of the groups of volunteers and patients at peak-systole, e-wave, and end-diastole; Figure S2: Relative error values of volume (a) and each hemodynamic parameter (b–g), obtained comparing the reference segmentation with segmentations given by erosion or dilation, for each group of volunteers and patients at e-wave. * indicates statistically significant differences ($p < 0.05$); Figure S3: Relative error values of volume (a) and each hemodynamic parameter (b–g), obtained comparing the reference segmentation with segmentations given by erosion or dilation, for each group of volunteers and patients at end-diastole. * indicates statistically significant differences ($p < 0.05$); Figure S4: Bullseye plots of mean hemodynamic parameters (a–f) across 16 segments for volunteers (i) and patients (ii) at the e-wave. * indicates statistically significant differences ($p < 0.05$).

Author Contributions: All authors were actively involved in reviewing and drafting the manuscript. All authors have approved the final version of this manuscript. P.F. developed the methodology to assess intracardiac flow, process all data, and perform statistical analysis and wrote the manuscript. J.S. developed the finite element method to evaluate the hemodynamic parameters and edited the manuscript. C.M. was the second observer in the inter-observer analysis and edited the manuscript. B.R. participated in the recruitment of patients, acquired CMR images, and edited the manuscript. E.K. developed the Eidolon software to register b-SSFP data and 4D Flow for the cohort of DCM patients and edited the manuscript. D.N. participated in the design of the study to recruit patients and edited the manuscript. J.M. participated helped develop the finite element method to assess hemodynamic parameters and edited the manuscript. D.H. participated in the development of the methodology to assess intracardiac hemodynamics and edited the manuscript. S.U. participated in the study's design, developing the method to assess intracardiac flow, helping with statistical analysis, and wrote the manuscript. All authors have read and agreed to the published version of the manuscript.

Funding: This work has been funded by projects PIA-ACT 192064 and the Millennium Nucleus on Cardiovascular Magnetic Resonance NCN17_129 of the Millennium Science Initiative, both of the National Agency for Research and Development, ANID. The authors also thanks to Fondecyt project 1181057 also by ANID. Franco P. thanks to ANID-PCHA/Doctorado-Nacional/2018-21180391. Sotelo J. thanks to CONICYT-FONDECYT Postdoctorado 2017 #3170737 and ANID-FONDECYT de Iniciación en Investigación #11200481.

Institutional Review Board Statement: All subjects participated under informed consent, with data collection approved by the Regional Ethics Committee, South East London, UK (REC, 12/LO/1456).

Informed Consent Statement: Informed consent was obtained from all subjects involved in the study.

Data Availability Statement: The datasets generated during and/or analyzed during the current study are not publicly available due to data privacy according to the rules of King's College London but could be available from Bram Ruijsink (b.ruijsink@gmail.com) on reasonable request.

Acknowledgments: This work has been funded by projects PIA-ACT 192064 and the Millennium Nucleus on Cardiovascular Magnetic Resonance NCN17_129 of the Millennium Science Initiative, both of the National Agency for Research and Development, ANID. The authors also give thanks to Fondecyt project 1181057, also by ANID. Franco P. gives thanks to ANID-PCHA/Doctorado-Nacional/2018-21180391. Sotelo J. gives thanks to CONICYT-FONDECYT Postdoctorado 2017 #3170737 and ANID-FONDECYT de Iniciación en Investigación #11200481.

Conflicts of Interest: The authors declare no conflict of interest.

Appendix A. Intra- and Inter-Observer Reproducibility

As shown in Figure A1, excellent intra-observer agreement with minimal mean differences and small limits of agreement were found for peak systole. Mean differences were: velocity magnitude -0.0003 ± 0.0118 m/s, kinetic energy $(-0.4580 \pm 0.7490) \times 10^{-9}$ J, vorticity magnitude 0.0029 ± 0.0249 1/s, helicity density $(-0.0978 \pm 0.5345) \times 10^{-3}$ m/s^2, viscous dissipation -0.0070 ± 0.0578 1/s^2, and energy loss $(-0.0124 \pm 0.2563) \times 10^{-9}$ W. Similar results were obtained at e-wave and end-diastole, as shown in Figures A2 and A3, respectively. Figure A4 demonstrates excellent inter-observer analysis agreement for peak systole. Mean differences were: velocity magnitude -0.0024 ± 0.0124 m/s, kinetic energy $(-0.0349 \pm 0.1093) \times 10^{-5}$ J, vorticity magnitude 0.1122 ± 0.7634 1/s, helicity density -0.0001 ± 0.0139 m/s^2, viscous dissipation -2.1652 ± 11.8205 1/s^2, and energy loss $(0.0012 \pm 0.3008) \times 10^{-8}$ W. Figures A5 and A6 show the results obtained at e-wave and end-diastole, respectively, with comparable results at peak systole.

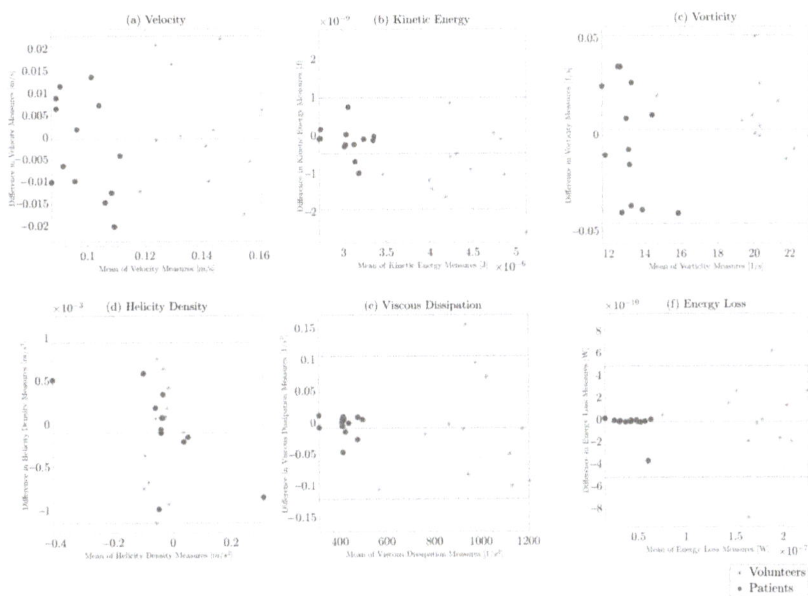

Figure A1. Bland-Altman plots represent the intra-observer reproducibility in the measurements of LV global hemodynamic parameters (**a–f**) at peak systole. The thick line represents the mean difference, and the thin lines represent the limits agreement (1.96 SD).

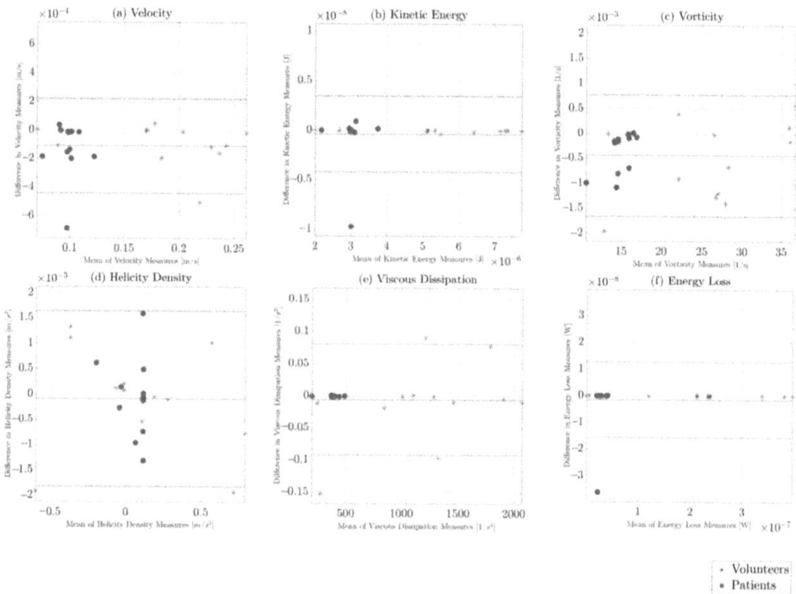

Figure A2. Bland-Altman plots represent the intra-observer reproducibility in the measurements of LV global hemodynamic parameters (**a–f**) at e-wave. The thick line represents the mean difference, and the thin lines represent the limits agreement (1.96 SD).

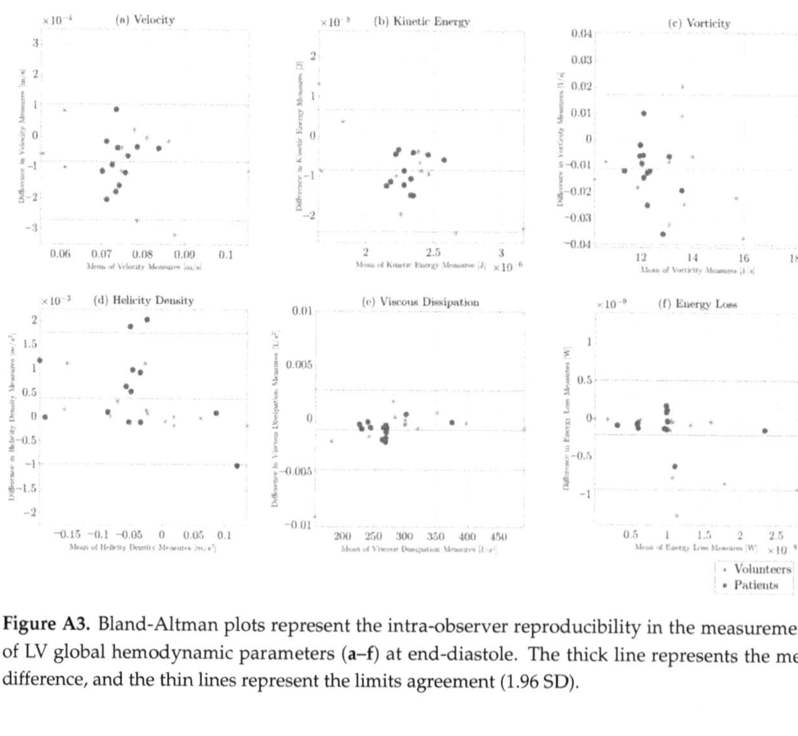

Figure A3. Bland-Altman plots represent the intra-observer reproducibility in the measurements of LV global hemodynamic parameters (**a–f**) at end-diastole. The thick line represents the mean difference, and the thin lines represent the limits agreement (1.96 SD).

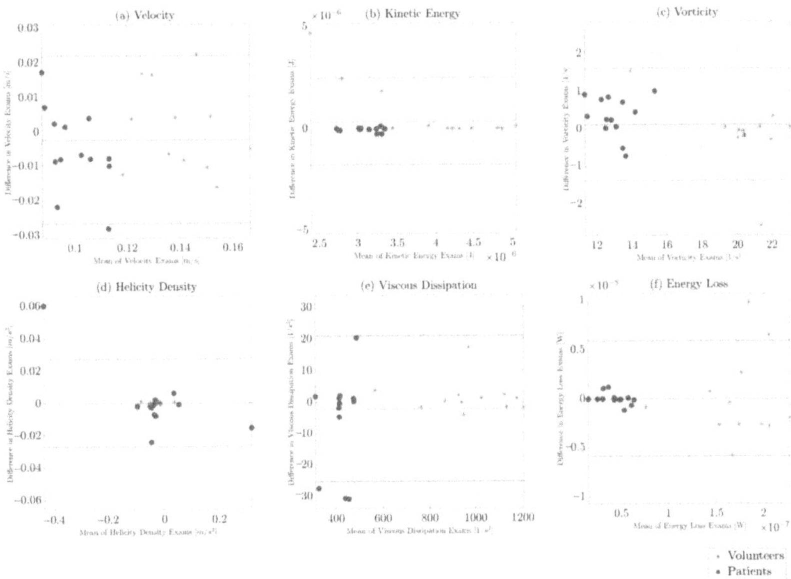

Figure A4. Bland-Altman plots represent the inter-observer reproducibility in the exams of LV global hemodynamic parameters (**a–f**) at peak systole. The thick line represents the mean difference, and the thin lines represent the limits agreement (1.96 SD).

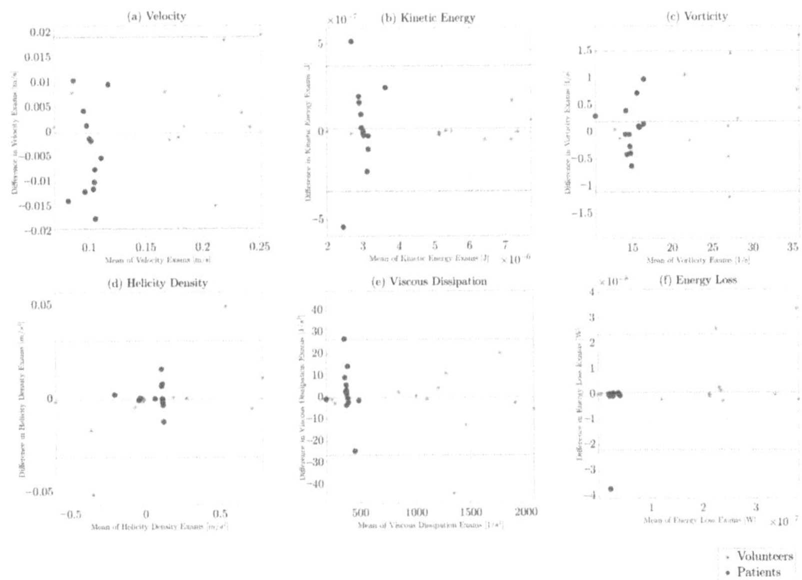

Figure A5. Bland-Altman plots represent the inter-observer reproducibility in the exams of LV global hemodynamic parameters (**a–f**) at e-wave. The thick line represents the mean difference, and the thin lines represent the limits agreement (1.96 SD).

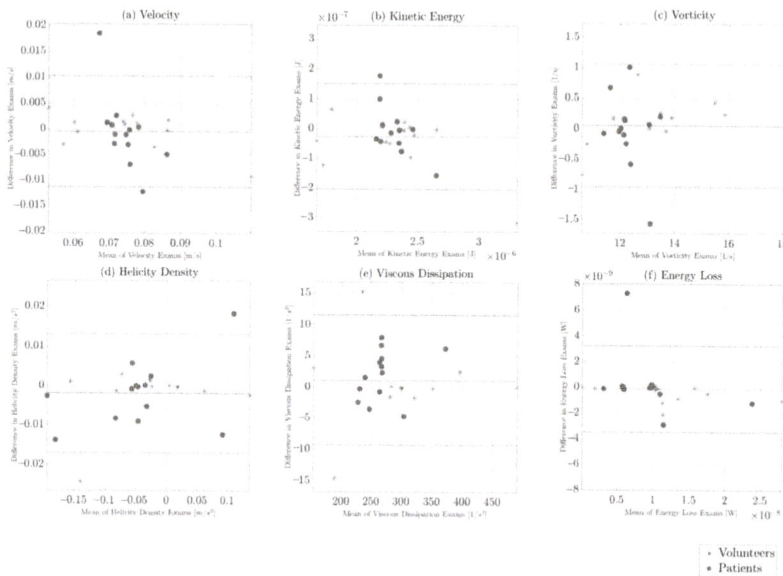

Figure A6. Bland-Altman plots represent the inter-observer reproducibility in the exams of LV global hemodynamic parameters (**a–f**) at end-diastole. The thick line represents the mean difference, and the thin lines represent the limits agreement (1.96 SD).

References

1. Hershberger, R.E.; Morales, A. Dilated Cardiomyopathy Overview. GeneReviews®. Available online: https://www.ncbi.nlm.nih.gov/books/NBK1309/ (accessed on 27 July 2007).
2. Mahmaljy, H.; Yelamanchili, V.S.; Singhal, M. Dilated Cardiomyopathy. In *StatPearls*; StatPearls Publishing: Treasure Island, FL, USA, 2021.
3. Hill, J.A.; Olson, E.N. Cardiac plasticity. *N. Engl. J. Med.* **2008**, *358*, 1370–1380. [CrossRef]
4. Wong, M.; Staszewsky, L.; Latini, R.; Barlera, S.; Glazer, R.; Aknay, N.; Hester, A.; Anand, I.; Cohn, J.N. Severity of left ventricular remodeling defines outcomes and response to therapy in heart failure: Valsartan heart failure trial (Val-HeFT) echocardiographic data. *J. Am. Coll. Cardiol.* **2004**, *43*, 2022–2027. [CrossRef]
5. McNally, E.M.; Mestroni, L. Dilated Cardiomyopathy: Genetic Determinants and Mechanisms. *Circ. Res.* **2007**, *121*, 731–748. [CrossRef]
6. Frangi, A.F.; Niessen, W.J.; Viergever, M.A. Three-dimensional modeling for functional analysis of cardiac images: A review. *IEEE Trans. Med. Imaging* **2007**, *20*, 2–25. [CrossRef]
7. Markl, M.; Frydrychowicz, A.; Kozerke, S.; Hope, M.; Wieben, O. 4D Flow MRI. *J. Magn. Reson. Imaging* **2012**, *36*, 1015–1036. [CrossRef] [PubMed]
8. Dyverfeldt, P.; Bissell, M.; Barker, A.J.; Bolger, A.F.; Carlhäll, C.-J.; Ebbers, T.; Francios, C.J.; Frydrychowicz, A.; Geiger, J.; Giese, D.; et al. 4D Flow cardiovascular magnetic resonance consensus statement. *J. Cardiovasc. Magn. Reson.* **2015**, *17*, 72. [CrossRef] [PubMed]
9. Sotelo, J.; Urbina, J.; Valverde, I.; Tejos, C.; Andia, M.; Hurtado, D.; Uribe, S. Three-dimensional quantification of vorticity and helicity from 3D cine PC-MRI using finite-element interpolations. *Magn. Reson. Med.* **2018**, *79*, 541–553. [CrossRef] [PubMed]
10. Sotelo, J.; Urbina, J.; Valverde, I.; Tejos, J.; Irarrazaval, P.; Andia, M.; Uribe, S.; Hurtado, D. 3D Quantification of Wall Shear Stress and Oscillatory Shear Index Using a Finite-Element Method in 3D CINE PC-MRI Data of the Thoracic Aorta. *IEEE Trans. Med. Imaging* **2016**, *35*, 1475–1487. [CrossRef]
11. Sotelo, J.; Dux-Santoy, L.; Guala, A.; Rodríguez-Palomares, J.; Evangelista, A.; Sing-Long, C.; Urbina, J.; Mura, J.; Hurtado, D.; Uribe, S. 3D axial and circumferential wall shear stress from 4D Flow MRI data using a finite element method and a laplacian approach. *Magn. Reson. Med.* **2018**, *79*, 2816–2823. [CrossRef]
12. Wong, K.K.; Kelso, R.M.; Worthley, S.G.; Sanders, P.; Mazumdar, J.; Abbott, D. Cardiac flow analysis applied to phase contrast magnetic resonance imaging of the heart. *Ann. Biomed. Eng.* **2009**, *37*, 1495–1515. [CrossRef]
13. Al-Wakeel, N.; Fernandes, J.F.; Amiri, A.; Siniawski, H.; Goubergrits, L.; Berger, F.; Kuehne, T. Hemodynamic and energetic aspects of the left ventricle in patients with mitral regurgitation before and after mitral valve surgery. *J. Magn. Reson. Imaging* **2015**, *42*, 1705–1712. [CrossRef] [PubMed]

14. Svalbring, E.; Fredriksson, A.; Eriksson, J.; Dyverfeldt, P.; Ebbers, T.; Bolder, A.F.; Engvall, J.; Carlhäll, C.-J. Altered Diastolic Flow Patterns and Kinetic Energy in Subtle Left Ventricular Remodeling and Dysfunction Detected by 4D Flow MRI. *PLoS ONE* **2016**, *11*, e0161391. [CrossRef]
15. Fredriksson, A.; Trzebiatowska-Krzynska, A.; Dyverfeldt, P.; Engvall, J.; Ebbers, T.; Carlhäll, C.-J. Turbulent kinetic energy in the right ventricle: Potential MR marker for risk stratification of adults with repaired Tetralogy of Fallot. *J. Magn. Reson. Imaging* **2018**, *47*, 1043–1053. [CrossRef] [PubMed]
16. Töger, J.; Kanski, M.; Carlsson, M.; Kovács, S.J.; Söderlind, G.; Arheden, H.; Heiberg, E. Vortex ring formation in the left ventricle of the heart: Analysis by 4D Flow MRI and Lagrangian coherent structures. *Ann. Biomed. Eng.* **2012**, *40*, 2652–2662. [CrossRef] [PubMed]
17. Browning, J.R.; Hertzberg, J.R.; Schroeder, J.D.; Fenster, B.E. 4D Flow Assessment of Vorticity in Right Ventricular Diastolic Dysfunction. *Bioengineering* **2017**, *4*, 30. [CrossRef]
18. Hirtler, D.; Garcia, J.; Barker, A.J.; Geiger, J. Assessment of intracardiac flow and vorticity in the right heart of patients after repair of tetralogy of Fallot by flow-sensitive 4D MRI. *Eur. Radiol.* **2016**, *26*, 3598–3607. [CrossRef] [PubMed]
19. Suwa, K.; Saitoh, T.; Takehara, Y.; Sano, M.; Saotome, M.; Urushida, T.; Katoh, H.; Satoh, H.; Sugiyama, M.; Wakayama, T.; et al. Intra-left ventricular flow dynamics in patients with preserved and impaired left ventricular function: Analysis with 3D cine phase contrast MRI (4D-Flow). *J. Magn. Reson. Imaging* **2016**, *44*, 1493–1503. [CrossRef]
20. Kanski, M.; Arvidsson, P.M.; Töger, J.; Borgquist, R.; Heiberg, E.; Carlsson, M.; Arheden, H. Left ventricular fluid kinetic energy time curves in heart failure from cardiovascular magnetic resonance 4D Flow data. *J. Cardiovasc. Magn. Reson.* **2015**, *17*, 111. [CrossRef]
21. Zajac, J.; Eriksson, J.; Dyverfeldt, P.; Bolder, A.F.; Carlhäll, C.-J. Turbulent kinetic energy in normal and myopathic left ventricles. *J. Magn. Reson. Imaging* **2015**, *41*, 1021–1029. [CrossRef]
22. Arvidsson, P.M.; Kovács, S.J.; Töger, J.; Borgquist, R.; Heiberg, E.; Carlsson, M.; Arheden, H. Vortex ring behavior provides the epigenetic blueprint for the human heart. *Sci. Rep.* **2016**, *6*, 22021. [CrossRef]
23. Pedrizzetti, G.; La Canna, G.; Alfieri, O.; Tonti, G. The vortex—An early predictor of cardiovascular outcome? *Nat. Rev. Cardiol.* **2014**, *11*, 545–553. [CrossRef]
24. Sjöberg, P.; Heiberg, E.; Wingren, P.; Johansson, J.R.; Malm, T.; Arheden, H.; Liuba, P.; Carlsson, M. Decreased Diastolic Ventricular Kinetic Energy in Young Patients with Fontan Circulation Demonstrated by Four-Dimensional Cardiac Magnetic Resonance Imaging. *Pediatr. Cardiol.* **2017**, *8*, 669–680. [CrossRef] [PubMed]
25. Sotelo, J.; Urbina, J.; Valverde, I.; Irarrázaval, P.; Hurtado, D.; Uribe, S. Quantification of wall shear stress using a finite-element method in multidimensional phase-contrast MR data of the thoracic aorta. *J. Biomech.* **2015**, *48*, 1817–1827. [CrossRef]
26. Elliott, P.; Andersson, B.; Arbustini, E.; Bilinska, Z.; Cecchi, F.; Charron, P.; Dubourg, O.; Kühl, U.; Maisch, M.; McKenna, W.J.; et al. Classification of the cardiomyopathies: A position statement from the European Society Of Cardiology Working Group on Myocardial and Pericardial Diseases. *Eur. Heart J.* **2008**, *29*, 270–276. [CrossRef] [PubMed]
27. Kerfoot, E.; Fovargue, L.; Rivolo, S.; Shi, W.; Rueckert, D.; Nordsletten, D.; Lee, J.; Chabiniok, R.; Razavi, R. Visualization and Computational Framework for Multi-modal Biomedical Data Analysis. In *Medical Imaging and Augmented Reality, MIAR*; Zheng, G., Liao, H., Jannin, P., Cattin, P., Lee, S.L., Eds.; Lecture Notes in Computer Science; Springer: Cham, Switzerland, 2016; Volume 9805.
28. Heiberg, E.; Sjögren, J.; Ugander, M.; Carlsson, M.; Engblom, H.; Arheden, H. Design and validation of Segment–freely available software for cardiovascular image analysis. *BMC Med. Imaging* **2010**, *10*, 1. [CrossRef]
29. Tufvesson, J.; Hedstrom, E.; Steding-Ehrenborg, K.; Carlsson, M.; Arheden, H.; Heiberg, E. Validation and Development of a New Automatic Algorithm for Time-Resolved Segmentation of the Left Ventricle in Magnetic Resonance Imaging. *Biomed. Res. Int.* **2015**, *2015*, 970357. [CrossRef] [PubMed]
30. Heiberg, E.; Wigström, L.; Carlsson, M.; Bolger, A.F.; Karlsson, M. Time Resolved Three-dimensional Segmentation of the Left Ventricle. *Proc. IEEE Comput. Cardiol.* **2005**, *32*, 599–602.
31. Fang, Q.; Boas, D. Tetrahedral mesh generation from volumetric binary and gray-scale images. In Proceedings of the IEEE International Symposium on Biomedical Imaging: From Nano to Macro, Boston, MA, USA, 28 June–1 July 2009; pp. 1142–1145.
32. Gu, Y.; Wang, C. A Study of Hierarchical Correlation Clustering for Scientific Volume Data. In *Advances in Visual Computing*; Bebis, G., Ed.; ISVC 2010. Lecture Notes in Computer Science; Springer: Berlin/Heidelberg, Germany, 2010; Volume 6455.
33. Lorenz, R.; Bock, J.; Barker, A.J.; von Knobelsdorff-Brenkenhoff, F.; Wallis, W.; Korvink, J.G.; Bissell, M.M.; Schulz-Menger, J.; Markl, M. 4D Flow MRI in bicuspid aortic valve disease demonstrates altered distribution of aortic blood flow helicity. *Magn. Reson. Med.* **2014**, *71*, 1542–1553. [CrossRef]
34. Fouras, A.; Soria, J. Accuracy of out-of-plane vorticity measurements derived from in-plane velocity field data. *Exp. Fluids* **1998**, *25*, 409–430. [CrossRef]
35. Amorim, S.; Rodrigues, J.; Campelo, M.; Moura, B.; Martins, E.; Macedo, F.; Silva-Cardoso, J.; Júlia Maciel, M. Left ventricular reverse remodeling in dilated cardiomyopathy-maintained subclinical myocardial systolic and diastolic dysfunction. *Int. J. Cardiovasc. Imaging* **2017**, *33*, 605–613. [CrossRef]
36. Friedberg, M.; Roche, S.; Mohammed, A.; Balasingam, M.; Atenafu, E.G.; Kantor, P.F. Left Ventricular Diastolic Mechanical Dyssynchrony and Associated Clinical Outcomes in Children With Dilated Cardiomyopathy. *Circ. Cardiovasc. Imaging* **2008**, *1*, 50–57. [CrossRef]

37. Dragulescu, A.; Mertens, L.; Friedberg, M. Interpretation of Left Ventricular Diastolic Dysfunction in Children With Cardiomyopathy by Echocardiography. *Circ. Cardiovasc. Imaging* **2013**, *6*, 254–261. [CrossRef]
38. Eriksson, J.; Bolger, A.; Ebbers, T.; Carlhäll, C.-J. Four-dimensional blood flow-specific markers of LV dysfunction in dilated cardiomyopathy. *Eur. Heart Cardiovasc. Imaging* **2013**, *14*, 417–424. [CrossRef] [PubMed]
39. Stoll, V.; Hess, A.; Rodgers, C.; Bissell, M.; Dyverfeldt, P.; Ebbers, T.; Myerson, S.G.; Carlhäll, C.-J.; Neubauer, S. Left Ventricular Flow Analysis: Novel Imaging Biomarkers and Predictors of Exercise Capacity in Heart Failure. *Circ. Cardiovasc. Imaging* **2019**, *12*, e008130. [CrossRef]
40. Eriksson, J.; Bolger, A.; Ebbers, T.; Carlhäll, C. Assessment of left ventricular hemodynamic forces in healthy subjects and patients with dilated cardiomyopathy using 4D Flow MRI. *Physiol. Rep.* **2016**, *4*, e12685. [CrossRef] [PubMed]
41. Foell, D.; Jung, B.A.; Germann, E.; Staehle, F.; Bode, C.; Henning, J.; Markl, M. Segmental Myocardial Velocities in Dilated Cardiomyopathy with and without Left Bundle Branch Block. *J. Magn. Reson. Imaging* **2013**, *37*, 119–126. [CrossRef] [PubMed]

applied sciences

MDPI

Article

4D Flow MRI in Ascending Aortic Aneurysms: Reproducibility of Hemodynamic Parameters

Joe F. Juffermans [1,*], Hans C. van Assen [1], Bastiaan J. C. te Kiefte [1], Mitch J. F. G. Ramaekers [1,2,3,4], Roel L. F. van der Palen [5], Pieter van den Boogaard [1], Bouke P. Adriaans [2,3,4], Joachim E. Wildberger [3,4], Ilona A. Dekkers [1], Arthur J. H. A. Scholte [6], Simon Schalla [2,3,4], Hildo J. Lamb [1] and Jos J. M. Westenberg [1]

1 Department of Radiology, Leiden University Medical Center, 2333 ZA Leiden, The Netherlands; h.c.van_assen@lumc.nl (H.C.v.A.); b.j.c.te_kiefte@lumc.nl (B.J.C.t.K.); mitch.ramaekers@mumc.nl (M.J.F.G.R.); p.j.van_den_boogaard@lumc.nl (P.v.d.B.); i.a.dekkers@lumc.nl (I.A.D.); h.j.lamb@lumc.nl (H.J.L.); j.j.m.westenberg@lumc.nl (J.J.M.W.)

2 Department of Cardiology, Maastricht University Medical Center, 6229 HX Maastricht, The Netherlands; bouke.adriaans@mumc.nl (B.P.A.); s.schalla@mumc.nl (S.S.)

3 Department of Radiology and Nuclear Medicine, Maastricht University Medical Center, 6229 HX Maastricht, The Netherlands; j.wildberger@mumc.nl

4 Cardiovascular Research Institute Maastricht, 6229 HX Maastricht, The Netherlands

5 Division of Pediatric Cardiology, Department of Pediatrics, Leiden University Medical Center, 2333 ZA Leiden, The Netherlands; r.vanderpalen@lumc.nl

6 Department of Cardiology, Leiden University Medical Center, 2333 ZA Leiden, The Netherlands; a.j.h.a.scholte@lumc.nl

* Correspondence: j.f.juffermans@lumc.nl

Abstract: (1) Background: Aorta hemodynamics have been associated with aortic remodeling, but the reproducibility of its assessment has been evaluated marginally in patients with thoracic aortic aneurysm (TAA). The current study evaluated intra- and interobserver reproducibility of 4D flow MRI-derived hemodynamic parameters (normalized flow displacement, flow jet angle, wall shear stress (WSS) magnitude, axial WSS, circumferential WSS, WSS angle, vorticity, helicity, and local normalized helicity (LNH)) in TAA patients; (2) Methods: The thoracic aorta of 20 patients was semi-automatically segmented on 4D flow MRI data in 5 systolic phases by 3 different observers. Each time-dependent segmentation was manually improved and partitioned into six anatomical segments. The hemodynamic parameters were quantified per phase and segment. The coefficient of variation (COV) and intraclass correlation coefficient (ICC) were calculated; (3) Results: A total of 2400 lumen segments were analyzed. The mean aneurysm diameter was 50.8 ± 2.7 mm. The intra- and interobserver analysis demonstrated a good reproducibility (COV = 16–30% and ICC = 0.84–0.94) for normalized flow displacement and jet angle, a very good-to-excellent reproducibility (COV = 3–26% and ICC = 0.87–1.00) for all WSS components, helicity and LNH, and an excellent reproducibility (COV = 3–10% and ICC = 0.96–1.00) for vorticity; (4) Conclusion: 4D flow MRI-derived hemodynamic parameters are reproducible within the thoracic aorta in TAA patients.

Keywords: aorta; aneurysm; hemodynamic; 4D flow MRI; reproducibility

Citation: Juffermans, J.F.; van Assen, H.C.; te Kiefte, B.J.C.; Ramaekers, M.J.F.G.; van der Palen, R.L.F.; van den Boogaard, P.; Adriaans, B.P.; Wildberger, J.E.; Dekkers, I.A.; Scholte, A.J.H.A.; et al. 4D Flow MRI in Ascending Aortic Aneurysms: Reproducibility of Hemodynamic Parameters. *Appl. Sci.* **2022**, *12*, 3912. https://doi.org/10.3390/app12083912

Academic Editor: Qi-Huang Zheng

Received: 18 March 2022
Accepted: 11 April 2022
Published: 13 April 2022

1. Introduction

Patients with a thoracic aortic aneurysm (TAA) have an increased risk for aortic rupture and dissection [1]. Clinical guidelines use a maximal cross-sectional diameter ≥55 mm as the main criterion for the recommendation of preemptive surgical aortic replacement [2,3]. However, the majority (60–96%) of dissections occur in aortas with diameters below this threshold [4,5]. Therefore, there is a need for additional and more sensitive markers to identify patients at high risk of progressive dilation or adverse aortic events [6].

Evaluation of aortic hemodynamics is considered to be highly promising for the prediction of progressive dilatation and adverse aortic events [7–14]; the blood flow over the cardiac cycle through the aorta can be analyzed using three-dimensional time-resolved phase-contrast magnetic resonance imaging (MRI), also known as four-dimensional (4D) flow MRI. From 4D flow MRI data, several relevant patient-specific hemodynamic parameters can be assessed. It has been demonstrated for patients with a tricuspid aortic valve (TAV) that the amount of flow displacement (a measure of eccentricity in the flow profile) is associated with aortic growth [7,8]. Furthermore, TAV patients showed decreased levels of vortical flow, helical flow, and wall shear stress (WSS) compared to healthy volunteers [9,10].

While such hemodynamic parameters could also be important for future clinical risk stratification of aneurysm patients, the majority of the studies assessing their agreements between observers and examinations only included healthy volunteers, patients with bicuspid aortic valves, or patients with mildly dilated aortas [13–22]. Hence, the reported outcomes are not fully translatable to TAA patients with substantially dilated aortas and a TAV. Therefore, the reproducibility of the hemodynamic parameters in such patients remains unknown [1]. Moreover, several previous studies quantified parameters based on (multiple) two-dimensional plane quantifications or ranked them qualitatively [15–18,21]. Thus, part of the spatial information is lost. In addition, the reproducibility of the WSS magnitude (WSS_{mag}) has been assessed well, but little is known about the reproducibility of the axial or circumferential WSS component (WSS_{ax}, WSS_{cir}, respectively) and the angle between both WSS components (WSS_{angle}).

To allow quantification of hemodynamic parameters in 4D flow MRI data, a cardiac phase-specific 3D lumen segmentation is required [23,24]. The shape of these lumen segmentations can be described by morphologic parameters such as the volume, centerline length, maximal diameter, and curvature radius. Interestingly, a recent study demonstrated that the curvature radius of the ascending aorta is associated with aortic growth [7].

In most commercial and research software tools, these lumen segmentations are constructed (semi-)automatically on the 4D flow MRI data and then manually adjusted when needed. The manual interaction may introduce observer-dependent variability in the quantitation of the hemodynamic parameters [25,26]. This observer-dependent variability of the 3D lumen segmentation can be assessed directly by evaluating the variability in morphologic parameters. The reproducibility of these morphologic parameters from 4D flow MRI data has been primarily reported in healthy volunteers [25].

Consequently, the aim of this study was to evaluate intra- and interobserver reproducibility of hemodynamic parameters, i.e., normalized flow displacement, flow jet angle, wall shear stress (magnitude, axial, circumferential, and angle), vorticity and helicity (absolute and local normalized) quantitatively assessed for six thoracic aortic lumen segments in patients with TAA and normally functioning TAV. In addition, intra- and interobserver reproducibility of morphological parameters, i.e., aortic volume, centerline length, lumen diameter, and curvature radius, were assessed.

2. Materials and Methods

2.1. Study Population

The study protocol was conducted in accordance with the Declaration of Helsinki and was approved by the local Medical Ethics Committee of the Leiden University Medical Center (G20.149, 9 October 2020), which waived the patient's informed consent for anonymized clinical data. The clinical database was used to identify TAA patients with a TAV who underwent a 4D flow MRI of the thoracic aorta between October 2018 and August 2021 (n = 260). Patients with aortic valve stenosis, regurgitation, or previous aortic surgery were excluded to obtain a selection of TAA patients without other pathologies, which potentially could also affect the reproducibility (n = 171). The clinical reports of the patients were evaluated for the maximal aortic diameter, and the 20 patients with the largest reported maximal diameters were included, see Figure 1. The presence of an aortic root or ascending aorta aneurysm was classified according to Della Corte's classification [27].

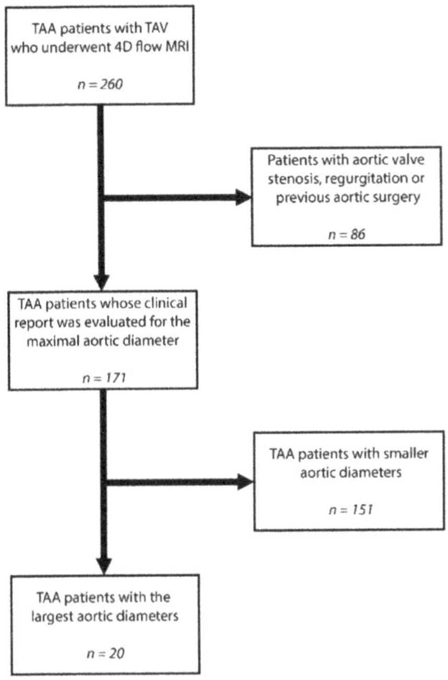

Figure 1. Population selection diagram. Abbreviations: TAA—thoracic aortic aneurysm, TAV—tricuspid aortic valve, MRI—magnetic resonance imaging, and 4D—four dimensional.

2.2. MRI Acquisition

The MRI examinations consisted of a 4D flow scan covering the entire thoracic aorta (for details, see Supplementary Materials Table S1). All subjects were scanned with a 3T-scanner (Ingenia or Elition, Philips Healthcare, Best, The Netherlands) using a FlexCoverage anterior and dStream Torso posterior coil. Concomitant gradient correction was performed using standard available scanner software. All data were visually inspected, and the absence of aliasing in the 4D flow MRI data was confirmed.

2.3. Image Analysis

The image analysis consisted of three parts: 4D flow MRI lumen segmentation, aorta morphology quantification, and hemodynamic quantification. First, the aortic lumen of the patients was segmented twice by the first observer (J.J.) and once by the second and third observers (M.R. and B.K., respectively). Lumen segmentation was performed using CAAS MR Solutions v5.2.1 (Pie Medical Imaging, Maastricht, The Netherlands). The software generated initial segmentations on the peak systolic phase and two consecutive phases before and after this peak systolic phase (i.e., five systolic phases, see Figure 2 and Supplementary Materials Video S1). Next, the aortic segmentations were manually corrected for all five phases and partitioned into six consecutive segments by manually placing anatomical planes perpendicular to the aortic centerline, respectively: the aortic root (from the aortic valve to the sinotubular junction), proximal ascending aorta (from the sinotubular junction to the mid-ascending aorta, defined as the midpoint on the centerline between the sinotubular junction and brachiocephalic artery), distal ascending aorta (from the mid-ascending aorta to the brachiocephalic artery), aortic arch (from the brachiocephalic artery until past the left subclavian artery), the proximal descending aorta (from the left subclavian artery to the mid-descending thoracic aorta, defined as the midpoint on the centerline between the left subclavian artery and the descending aorta at the level of the aortic valve), and distal descending aorta (from the mid-descending thoracic aorta to the descending aorta at the level of the aortic valve). More details about the aortic segmentation are described by van der Palen et al. [20]. In addition, the stroke volume and cardiac output were assessed at the level of the sinotubular junction by the first observer for characterizing the left ventricular function of the population using CAAS MR Solutions.

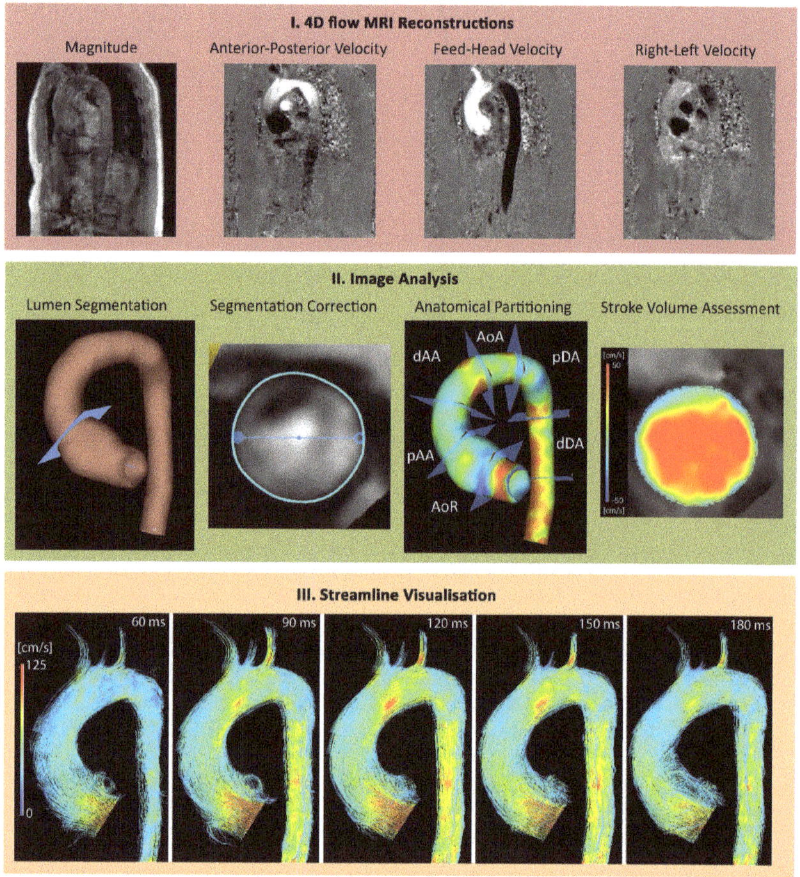

Figure 2. 4D flow MRI visualization using CAAS MR Solutions v.5.2.1. Example of a patient (male, 45 years old) with aortic root aneurysm with a diameter of 52 mm. (**I**) 4D flow MRI reconstructions. (**II**) Image analysis including the lumen segmentation, the manual segmentation correction, anatomical partitioning visualized on the wall shear stress maps, and stroke volume assessment. (**III**) Streamline visualization of the 4D flow MRI at the 5 systolic phases (60, 90, 120, 150, and 180 ms). Abbrevations: AoR—aortic root, pAA—proximal ascending aorta, dAA—distal ascending aorta, AoA—aortic arch, pDA—proximal descending aorta, and dDA—distal descending aorta. Secondly, the morphology of the thoracic aortas was analyzed using an in-house developed Python software (Python Software Foundation, Wilmington, DE, USA). This tool was used to quantify aortic volume, centerline length, maximal diameter, and curvature radius of each anatomical segment and for each of the five systolic phases. The aortic diameter was determined by first constructing a cross-section perpendicular to the centerline at every millimeter. Next, at each cross-section, the maximal radial spike length was calculated. The curvature radius was derived by fitting a circle through the segments' centerline as previously described [25]. The maximal aortic diameter within each segment and the radius of the fitted circle were used for statistical analysis.

Thirdly, the hemodynamics of the thoracic aorta were quantified using methods described in previous studies [13,15,19]. These methods were implemented in an in-house developed Python software [9,25]. Initially, the mean blood velocity over all subjects was quantified per anatomical segment. The tool was then used to quantify the normalized flow displacement, flow jet angle, vorticity norm, absolute helicity, and absolute LNH for each anatomical segment and each systolic phase. The normalized flow displacement and jet angle were determined by first constructing a cross-section perpendicular to the centerline at every millimeter. Next, at each cross-section, the normalized flow displacement (distance between the centerline point and center of velocity normalized for the vessel diameter times 100%) and flow jet angle (angle between the centerline and mean velocity vector) were calculated as previously described [15]. The maximal normalized flow displacement and flow jet angle within each segment were assessed and used for statistical analysis. For the calculation of vorticity and helicity of the velocity vector fields ($\underset{V}{\rightarrow}$), a spatial Gaussian differential operator ($\underset{\nabla}{\rightarrow}$) with a standard deviation equal to the reconstructed in-plane voxel size was used. The vorticity ($\underset{\omega}{\rightarrow} = \underset{\nabla}{\rightarrow} \times \underset{V}{\rightarrow}$) norm and absolute helicity ($H = |\underset{V}{\rightarrow} \cdot \underset{\omega}{\rightarrow}|$) were quantified via methods previously described by Ramaekers et al. [9]. The vorticity norm and absolute helicity were both normalized for the reconstructed voxel size, and the mean values within each segment were used for statistical analysis. Since a previous study demonstrated that a threshold of >0.6 for absolute LNH ($LNH = \frac{\vec{V} \cdot \vec{\omega}}{|\underset{V}{\rightarrow}||\underset{\omega}{\rightarrow}|}$) is the optimal criterion for detecting differences between patients and healthy volunteers [13], this threshold was used in our study to select volumes with increased absolute LNH. The volumes with increased absolute LNH were assessed and used for the statistical analysis. The in-house developed software can be made available for scientific collaboration per reasonable request. To derive 3D WSS maps for each of the five systolic phases, CAAS MR Solutions v5.2 was used. At each point of the 3D WSS map, WSS_{mag} was calculated as well as the WSS_{ax} and WSS_{cir} components. The WSS_{angle} between the WSS_{cir} and WSS_{ax} was quantified for each point as previously described [19]. For each anatomical segment, the mean WSS_{mag}, WSS_{ax}, WSS_{cir}, and WSS_{angle} were used for statistical analysis.

2.4. Observer Training

All observers were acquainted with CAAS MR Solutions software. The first, second, and third observer had 4, 3, and 1 years' experience with lumen segmentation on 4D flow MRI data in TAA patients, respectively. Since the observers were recruited at two clinical centers and observers had different levels of experience, all observers received feedback on the segmented aorta from an experienced cardiovascular MRI researcher (J.J.M.W. with >15 years of 4D flow MRI experience) after analyzing the training cohort (n = 10 TAA patients). Hereafter, the results of the training cohort were not improved. Next, the observers analyzed the validation cohort (n = 10 TAA patients).

2.5. Statistical Analysis

Statistical analysis was performed using the open-source SciPy v1.2.1. software [28]. Parametric and non-parametric data are expressed as mean ± standard deviation (SD) or median (Q1–Q3), respectively. The Shapiro–Wilk test was used to verify the normality of the data. To assess the intra- and interobserver reproducibility, Bland–Altman analysis was performed, and the coefficient of variation (COV), Spearman rank correlation coefficient (r), and intraclass correlation coefficient (ICC) (two-way mixed effects, absolute agreement) were calculated. For the Bland–Altman analysis, the mean difference and limits of agreement (LoA = 1.96 × SD) were computed [29]. In addition, the relative difference in LoA between the training and validation cohort was calculated (LoA_{diff} = ((validation LoA–training LoA)/training LoA) × 100%). The COV was classified as low (≤10%), intermediate (11–20%), high (21–30%), or very high (>30%). The r and ICC were classified as poor (<0.50), moderate (0.50–0.69), good (0.70–0.84), very good (0.85–0.94), or excellent (≥0.95) [25]. A *p*-value of <0.05 was considered statistically significant.

3. Results

The patient characteristics are presented in Table 1. In Table 2, median values of hemodynamic and morphologic parameters are presented. Of the selected TAA patients, 13 presented with an aneurysm in the aortic root and 7 with an ascending aorta aneurysm. The mean TAA diameter and mean blood velocity were 50.8 ± 2.7 mm and 42.5 ± 14.9 cm/s, respectively.

Table 1. Population characteristics.

Characteristic	Quantity
Population size	20
Male (%)	18 (90%)
Age (years)	53 ± 14
Height (cm)	186 ± 8
Weight (kg)	90 ± 11
TAA diameter (mm)	50.8 ± 2.7
Heart rate (bpm)	68 ± 11
Stroke Volume (mL)	105 ± 24
Cardiac Output (L min^{-1})	7.0 ± 1.8
Systolic blood pressure (mm Hg)	132 ± 14
Diastolic blood pressure (mm Hg)	81 ± 7
Trigger delay peak systole—2 phases (ms)	102 ± 31
Trigger delay peak systole—1 phase (ms)	133 ± 32
Trigger delay peak systole (ms)	163 ± 34
Trigger delay peak systole + 1 phase (ms)	194 ± 35
Trigger delay peak systole + 2 phases (ms)	224 ± 36

Data notated as the mean ± standard deviation and presented over all subjects. Abbreviation: TAA—thoracic aortic aneurysm.

Table 2. The hemodynamic and morphologic characteristics assessed in the thoracic aorta.

Parameter	Anatomical Segments					
	AoR	pAA	dAA	AoA	pDA	dDA
Flow Displacement (%)	7.5 (5.3–10.2)	9.8 (7.1–16.2)	6.3 (4.9–9.0)	5.7 (4.7–6.5)	6.5 (5.8–7.6)	4.7 (4.1–5.7)
Flow Jet Angle (°)	17.6 (13.7–21.9)	17.6 (11.9–27.0)	14.8 (10.2–21.3)	14.6 (12.5–17.3)	15.7 (13.1–18.4)	9.1 (7.6–10.6)
WSS magnitude (mPa)	561 (483–654)	763 (570–896)	750 (595–955)	746 (560–992)	877 (706–1112)	939 (785–1179)
Axial WSS (mPa)	371.9 (335–478)	627.9 (432–734)	702.6 (551–893)	700.6 (525–949)	801.5 (670–1045)	915.2 (748–1145)
Circumferential WSS (mPa)	331 (286–410)	331 (247–405)	204 (159–262)	182 (153–215)	243 (180–296)	205 (147–244)
WSS angle (°)	42.5 (37.4–45.7)	32.0 (26.8–37.1)	18.3 (13.7–24.0)	14.0 (12.1–16.9)	16.8 (14.4–20.2)	11.7 (9.6–14.4)
Vorticity Norm (s^{-1}·mL^{-1})	18,882 (16,767–21,131)	12,909 (10,629–15,526)	10,390 (7695–12,407)	11,338 (9273–14,205)	14,422 (11,800–19,372)	14,150 (11,164–17,990)
Absolute Helicity (m·s^{-2}·mL^{-1})	3084 (2327–3603)	1997 (1287–2729)	1098 (624–1544)	1207 (950–1749)	1873 (1149–2990)	1464 (1059–2263)

Table 2. *Cont.*

Parameter	Anatomical Segments					
	AoR	pAA	dAA	AoA	pDA	dDA
Absolute Local Normalized Helicity Volume (mL)	19.4 (10.5–27.3)	24.8 (18.9–30.2)	15.2 (10.4–21.7)	9.4 (7.0–13.6)	15.6 (12.2–18.2)	17.7 (13.7–19.6)
Volume (mL)	44.1 (25.2–59.8)	53.2 (36.9–59.7)	41.1 (31.0–53.8)	28.6 (20.0–31.5)	38.4 (29.9–45.2)	31.5 (24.2–38.0)
Centerline Length (mm)	31.0 (24.3–37.1)	40.8 (36.6–44.0)	40.0 (36.7–44.3)	37.3 (34.0–38.7)	65.4 (57.0–69.2)	64.1 (58.0–68.9)
Maximal Diameter (mm)	48.7 (40.0–51.3)	44.7 (40.4–46.6)	41.2 (35.4–44.9)	34.3 (31.5–36.3)	31.3 (28.2–32.6)	27.3 (24.9–29.7)
Curvature Radius (mm)	51.4 (42.9–71.1)	38.5 (36.2–46.6)	50.6 (43.3–61.3)	46.1 (40.8–59.7)	45.5 (37.3–53.7)	138.1 (102.0–164.5)

Data notated as the median (Q1–Q3) and presented over all subject and systolic phases. Abbreviations: AoR—aortic root, pAA—proximal ascending aorta, dAA—distal ascending aorta, AoR—aortic arch, pDA—proximal descending aorta, and dDA—distal descending aorta.

A total of 2400 aorta segments were analyzed for the intra- and interobserver reproducibility analysis, see Figure 3. The result of these analyses over all anatomical segments and systolic phases for the validation and training cohort are presented in Table 3 and Supplementary Materials Table S2, respectively.

Table 3. Intra- and interobserver reproducibility: results of the validation cohort.

Study	Bland–Altman		COV (%)	Spearman RankCorrelation Coefficient	Intraclass Correlation Coefficient	LoA_{diff} (%)
	Mean Diff	LoA				
			Flow Displacement (%)			
IA-O	−0.1	2.8	19	0.81	0.94	−9
IE-O1	−0.5	4.8	32	0.70	0.84	15
IE-O2	0.1	4.4	30	0.70	0.85	3
IE-O3	0.6	4.1	27	0.79	0.87	7
			Flow Jet Angle (°)			
IA-O	0.3	5.0	16	0.92	0.93	−16
IE-O1	−0.3	8.0	25	0.80	0.83	−4
IE-O2	0.2	7.1	23	0.85	0.86	−9
IE-O3	0.5	8.7	27	0.78	0.80	−4
			WSS Magnitude (mPa)			
IA-O	0	43	3	1.00	1.00	−15
IE-O1	−33	232	14	0.90	0.93	−10
IE-O2	17	161	10	0.93	0.97	−25
IE-O3	50	242	15	0.89	0.93	30
			Axial WSS (mPa)			
IA-O	−0	47	3	1.00	1.00	−59
IE-O1	−30	240	16	0.92	0.94	−58
IE-O2	19	163	11	0.95	0.97	−51
IE-O3	49	244	17	0.92	0.94	51

Table 3. *Cont.*

Study	Bland–Altman		COV (%)	Spearman RankCorrelation Coefficient	Intraclass Correlation Coefficient	LoA_{diff} (%)
	Mean Diff	LoA				
			Circumferential WSS (mPa)			
IA-O	−1	39	8	0.98	0.98	11
IE-O1	0	96	19	0.89	0.87	−18
IE-O2	3	73	15	0.93	0.93	0
IE-O3	3	103	20	0.86	0.87	−23
			WSS Angle (°)			
IA-O	0.1	3.3	7	0.99	0.99	−47
IE-O1	0.2	8.9	19	0.89	0.94	−8
IE-O2	−0.7	7.5	16	0.95	0.96	−21
IE-O3	−0.9	8.7	18	0.91	0.95	5
			Vorticity Norm ($s^{-1}\cdot mL^{-1}$)			
IA-O	37	860	3	1.00	1.00	−40
IE-O1	180	2001	7	0.97	0.98	−4
IE-O2	93	2485	9	0.96	0.97	−22
IE-O3	−87	2779	10	0.96	0.96	6
			Absolute Helicity ($m\cdot s^{-2}\cdot mL^{-1}$)			
IA-O	30	223	6	1.00	1.00	−9
IE-O1	−16	394	10	0.98	0.98	14
IE-O2	17	603	15	0.96	0.96	−14
IE-O3	32	631	16	0.96	0.96	21
			Absolute Local Normalized Helicity Volume (mL)			
IA-O	0.0	3.3	10	0.99	0.98	−18
IE-O1	0.2	5.9	18	0.95	0.92	6
IE-O2	0.9	8.4	26	0.88	0.86	−28
IE-O3	0.7	6.3	20	0.92	0.91	−10
			Volume (mL)			
IA-O	0.2	6.8	9	0.98	0.98	4
IE-O1	1.0	12.7	18	0.93	0.90	−21
IE-O2	1.9	15.7	22	0.89	0.86	−13
IE-O3	0.9	10.8	15	0.94	0.92	−41
			Centerline Length (mm)			
IA-O	0.9	7.1	8	0.94	0.97	−7
IE-O1	0.2	11.9	14	0.90	0.92	−36
IE-O2	3.5	12.2	14	0.89	0.92	−20
IE-O3	3.3	8.0	9	0.94	0.96	−53
			Maximal Diameter (mm)			
IA-O	−0.2	1.9	3	0.99	0.99	−14
IE-O1	−0.0	4.0	6	0.98	0.97	−3
IE-O2	−0.9	5.0	7	0.98	0.96	−42
IE-O3	−0.9	4.3	6	0.98	0.97	−36

Table 3. *Cont.*

Study	Bland–Altman		COV (%)	Spearman RankCorrelation Coefficient	Intraclass Correlation Coefficient	LoA_{diff} (%)
	Mean Diff	LoA				
			Curvature Radius (mm)			
IA-O	2.3	37.2	29	0.87	0.92	30
IE-O1	−2.7	46.4	35	0.77	0.90	14
IE-O2	3.6	42.4	34	0.77	0.90	−72
IE-O3	6.3	43.5	34	0.72	0.92	−70

Characteristics are presented per cohort over all subjects, anatomical segments, and systolic phases (n = 300). All probability values <0.01. Abbreviations: Mean Diff—mean difference; LoA—limits of agreement (1.96× standard deviation); COV—coefficient of variation; IA-O—intraobserver analysis by the first observer with 4 years' experience; IE-O1—interobserver analysis by the first and second observer with 4 and 3 years' experience, respectively; IE-O2—interobserver analysis by the first and third observer with 4 and 1 years' experience, respectively; IE-O3—interobserver analysis by the second and third observer with 3 and 1 years' experience, respectively; WSS—wall shear stress; and LoA_{diff} –relative difference in limits of agreement between the training and validation cohort.

Intra- and interobserver analysis of the hemodynamic parameters demonstrated a good reproducibility in the validation cohort for normalized flow displacement and flow jet angle (LoA ≤4.8% and ≤8.7° respectively, COV = 16–32%, r = 0.70–0.92, ICC = 0.80–0.94), an excellent reproducibility for vorticity norm (LoA ≤2779 $s^{-1}{\cdot}mL^{-1}$, COV = 3–10%, r = 0.96–1.00, ICC = 0.96–1.00), and a very good reproducibility for absolute WSS_{mag}, WSS_{ax}, WSS_{cir}, WSS_{angle}, absolute helicity and volumes with increased absolute LNH (LoA ≤242 mPa, ≤244 mPa, ≤103 mPa, ≤8.9°, ≤631 $m{\cdot}s^{-2}{\cdot}mL^{-1}$ and ≤8.4 mL, respectively; COV = 3–26%, r = 0.86–1.00, ICC = 0.86–1.00). See Supplementary Materials Figures S1–S9 for the corresponding Bland–Altman plots. These plots demonstrated no striking dependency of the measured differences relative to the mean for the analyses and hemodynamic parameters.

Intra- and interobserver analysis of the morphologic parameters demonstrated a very good reproducibility in the validation cohort for volume and centerline length (LoA ≤15.7 mL and ≤12.2 mm, respectively; COV = 8–22%, r = 0.89–0.98, ICC = 0.90–0.98), and an excellent reproducibility for diameter (LoA ≤5.0 mm, COV = 3–7%, r = 0.98–0.99, ICC = 0.96–0.99) and a good reproducibility for curvature radius (LoA ≤46.4 mm, COV = 21–35%, r = 0.72–0.87, ICC = 0.90–0.92). See Supplementary Materials Figures S10–S13 for the corresponding Bland–Altman plots. These plots demonstrated no dependency of the measured differences relative to the mean for most analyses and morphologic parameters, except for the maximal diameter, which had a slightly increasing difference relative to the mean for the interobserver analyses.

The intra- and interobserver reproducibility results per anatomical segment are presented in Supplementary Materials Tables S3 and S4, respectively. The intra- and interobserver analysis showed for the validation cohort a similar reproducibility for most morphologic and hemodynamic parameters per segment. However, the reproducibility was slightly decreased within the aortic root for several parameters (i.e., flow displacement, WSS_{mag}, WSS_{ax}, absolute helicity, volumes with increased absolute LNH, aortic segment volume, centerline length, and curvature radius). Furthermore, the reproducibility of the maximal diameter was slightly decreased within the proximal part of the aorta (i.e., the aortic root, proximal and distal ascending aorta) compared to the other anatomical segments further distally to the ascending aorta (intraobserver LoA ≤2.7 and ≤1.8 mm; interobserver LoA ≤7.6 and ≤3.3 mm, respectively). Moreover, the reproducibility of WSS_{angle} was slightly decreased within the proximal and distal descending aorta.

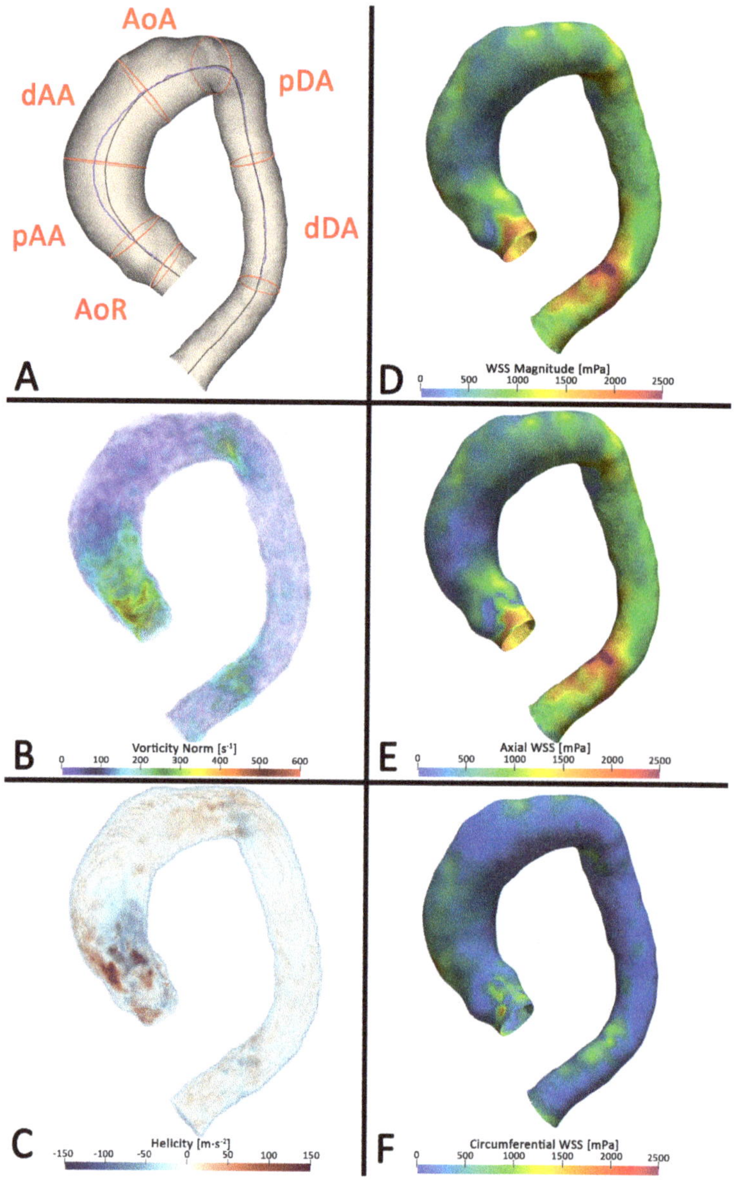

Figure 3. Aortic segments and hemodynamic parameters. Example of a patient (male, 71 years old) with an ascending aorta aneurysm with a diameter of 48 mm. (**A**) Anatomical segments with centerline (black color) and flow displacement line (blue color). (**B**) Vorticity norm. (**C**) Absolute helicity. Positive and negative magnitudes indicate a clockwise and anti-clockwise rotation, respectively. (**D**) Wall shear stress magnitude. (**E**) Axial wall shear stress. (**F**) Circumferential wall shear stress. Abbreviations: AoR—aortic root, pAA—proximal ascending aorta, dAA—distal ascending aorta, AoR—aortic arch, pDA—proximal descending aorta, dDA—distal descending aorta, and WSS—wall shear stress.

Compared to the training cohort, the validation cohort demonstrated reduced LoA over all analyses for morphologic and hemodynamic parameters (LoA_{diff} = −12 [−26–3]%). This improvement was more evident for the morphologic parameters compared to the hemodynamic parameters (LoA_{diff} = −20 [−41–−6]% and −9 [−21–5]%, respectively). Moreover, this improvement was also more evident within the aortic root, proximal and distal ascending aorta compared to the aortic arch, proximal and distal descending aorta (LoA_{diff} = −10 [−35–11] % and 2 [−18–41]%, respectively).

4. Discussion

The main findings of the current study are: (1) hemodynamic parameters (i.e., normalized flow displacement, flow jet angle, WSS_{mag}, WSS_{ax}, WSS_{cir}, WSS_{angle}, vorticity norm, absolute helicity, and volumes with increased absolute LNH) and morphological parameters (volume, centerline length, diameter and curvature radius) can be assessed with very good reproducibility in the thoracic aorta of TAA patients with a TAV, (2) the reproducibility of these parameters is slightly decreased within the aortic root, and (3) the reproducibility is affected by the observer's experience.

The intra- and interobserver analyses generally demonstrated a very good reproducibility for the assessment of hemodynamic parameters in TAA patients with a TAV. Similar results have been described previously for flow displacement, flow jet angle, WSS_{mag}, WSS_{angle}, secondary flow patterns (which are reflected in vorticity and helicity), and volumes with increased absolute LNH in healthy volunteers and in patients with none-to-marginally dilated aortas [11–19,30], but not yet in a clinically relevant patient group, such as patients with TAA. Moreover, the Bland–Altman plots of the current study demonstrated for most hemodynamic parameters and analyses no striking dependency of the measured differences relative to the mean.

Furthermore, the intra- and interobserver analyses demonstrated very good reproducibility of the morphologic parameters in TAA patients with a TAV. Similar reproducibility results have been described previously for volume, centerline length, maximal diameter, and curvature radius in healthy volunteers [25]. However, the current study demonstrated wider LoAs for all morphologic parameters within TAA patients compared to healthy volunteers. These wider LoAs in TAA patients may potentially be associated with the presence of dilation since, for a constant flow rate, the blood velocity is inversely related to the vessel's diameter. Therefore, a larger area of reduced velocity-to-noise ratio is expected for TAA patients compared to healthy volunteers, especially close to the vessel wall. This reduced velocity-to-noise ratio potentially can have a negative influence on the reproducibility of aortic lumen segmentation. However, the velocity-to-noise ratio within these areas may be improved by 4D flow MRI sequences with multiple-velocity-encoding gradients, which consequently may also improve the reproducibility of hemodynamics parameters [31,32].

When analyzing the results per anatomical segment, it was observed that the reproducibility was slightly reduced but still acceptable for most hemodynamic and morphologic parameters within the aortic root. This slight reduction in reproducibility was also seen in the proximal and distal ascending aorta for the maximal diameter. This reduced reproducibility is potentially introduced by the movement of the aortic root and ascending aorta in contrast to the relatively fixed-positioned aortic arch and descending aorta [33–35]. This effect related to the cardiac movement (contraction and relaxation) may further be amplified by the heart rate variation and breathing variation during the acquisition. Alongside, inadequate respiratory compensation or inadequate electrocardiographic gating may also perturbate the 4D flow MRI data. Moreover, within the aortic root, complex secondary flow patterns (recirculating flow) are present [9,36]. Since this recirculating flow has a relatively low velocity compared to the velocity-encoding sensitivity, these recirculating areas have a reduced velocity-to-noise ratio close to the vessel wall and potentially introduce more observer-dependent variability on the aortic lumen segmentation.

The observer-dependent variability of the 3D lumen segmentation was assessed directly by evaluating the variability in morphologic parameters. This analysis also demonstrated, for example, wider LoAs for the maximal diameter within the proximal part of the aorta (i.e., aortic root, proximal and distal ascending aorta) compared to thoracic aorta distal to the ascending part (intraobserver LoA ≤2.7 and ≤1.8 mm; interobserver LoA ≤7.6 and ≤3.3 mm, respectively). In contrast, a recent study assessing the interobserver reproducibility of manually measured diameters imaged with computed tomography demonstrated in the aortic root, ascending and descending aorta an LoA of 2.5, 2.1, and 1.9 mm, respectively [37]. For MRI, it is recommended to measure the maximal diameter on an MRI sequence with an isotropic voxel size ≤1.5 mm to achieve a clinically acceptable accuracy [38]. Since the 4D flow MRI sequence applied in the current study had an isotropic spatial resolution of 2.5–3.0 mm, the morphologic parameters derived from this sequence should not be used to define the maximal aortic diameter of TAA patients. However, for the quantification of hemodynamic parameters, an isotropic resolution of ≤3.0 mm is indicated as sufficient for the 4D flow MRI [23]. Of note, the 4D flow MRI was acquired to assess the patient's aorta hemodynamics and therefore was not meant to describe the patient's aorta morphology (i.e., TAA diameter).

The decreased LoA within the validation cohort compared to the training cohort indicate that the reproducibility likely is affected by the observer's experience, especially for the aortic root, proximal, and distal ascending aorta. Moreover, the observer's experience likely plays a larger role in the assessment of morphologic parameters compared to the hemodynamic parameters, considering the decrease in LoA between the training and validation cohort. This result demonstrates the robustness of the image analysis for the quantification of hemodynamic parameters, despite the observer-dependent variability and experience.

No major limitations were observed for the hemodynamics quantification in the segments of the thoracic aorta. Although no clear reproducibility limitations were observed for TAA patients with a TAV, still some variation is present in the hemodynamic parameters. This variation must be taken into account when these hemodynamic parameters are used for the prediction of progressive dilation and adverse aortic events in TAA patients with a TAV.

A limitation of the present study was the small population size. However, the robustness of the study was improved by including multiple observers, analyzing five systolic phases, and dividing the thoracic aorta into six anatomical segments. This resulted in a total of 2400 lumen segments for which the hemodynamic and morphologic parameters were quantified. Moreover, future research should include patients with other aortic pathologies. Finally, for patients with TAA, no data were available on repeated 4D flow MRI acquisition; thus, interexamination reproducibility could not be assessed.

5. Conclusions

The current study demonstrated a very good reproducibility for all hemodynamic parameters within the thoracic aorta in TAA patients. This allows precise quantification of the patient's aorta hemodynamics from a 4D flow MRI.

Supplementary Materials: The following supporting information can be downloaded at: https://www.mdpi.com/article/10.3390/app12083912/s1, Figure S1: Bland–Altman plots for the normalized flow displacement; Figure S2: Bland–Altman plots for the flow jet angle; Figure S3: Bland–Altman plots for the wall shear stress magnitude; Figure S4: Bland–Altman plots for the axial wall shear stress; Figure S5: Bland–Altman plots for the circumferential wall shear stress; Figure S6: Bland–Altman plots for the wall shear stress angle; Figure S7: Bland–Altman plots for the vorticity norm; Figure S8: Bland–Altman plots for the absolute helicity; Figure S9: Bland–Altman plots for the absolute local normalized helicity volume; Figure S10: Bland–Altman plots for the volume; Figure S11: Bland–Altman plots for the centerline length; Figure S12: Bland–Altman plots for the maximal diameter; Figure S13: Bland–Altman plots for the curvature radius; Table S1: Characteristics of the 4D flow MRI Sequence; Table S2: Intra- and interobserver reproducibility: results of the training cohort; Table S3: Intraobserver reproducibility: results of both cohorts per anatomical segment; Table S4: Interobserver reproducibility: results of both cohorts per anatomical segment; Video S1: Pathline visualization of the 4D flow MRI of a patient with an aortic root aneurysm with a diameter of 52 mm.

Author Contributions: All authors were actively involved in reviewing and editing the manuscript. J.F.J. wrote the original draft version of the manuscript; J.F.J., B.J.C.t.K., R.L.F.v.d.P., P.v.d.B., I.A.D., A.J.H.A.S., H.J.L. and J.J.M.W. contributed to the data curation; J.F.J. and H.C.v.A. developed the in-house developed software; J.F.J., M.J.F.G.R. and B.J.C.t.K. performed the image analysis under supervision of J.J.M.W.; H.J.L., J.J.M.W., A.J.H.A.S., S.S., J.E.W. and B.P.A. contributed to the project administration and funding acquisition. All authors have read and agreed to the published version of the manuscript.

Funding: This work was funded by the Dutch Heart Foundation (CVON2017-08-RADAR).

Institutional Review Board Statement: The study was conducted in accordance with the Declaration of Helsinki and approved by the Institutional Ethics Committee of Leiden University Medical Center (project code G20.149, 9 October 2020).

Informed Consent Statement: Patient consent was waived due to the retrospective inclusion of anonymized clinical data. Patients were not subjected to any research procedures and had the option to withdraw their participation.

Data Availability Statement: Data can be made available upon reasonable request.

Conflicts of Interest: The authors declare no conflict of interest.

References

1. Davies, R.; Gallo, A.; Coady, M.A.; Tellides, G.; Botta, D.M.; Burke, B.; Coe, M.P.; Kopf, G.S.; Elefteriades, J.A. Novel Measurement of Relative Aortic Size Predicts Rupture of Thoracic Aortic Aneurysms. *Ann. Thorac. Surg.* **2006**, *81*, 169–177. [CrossRef] [PubMed]
2. Hiratzka, L.F.; Bakris, G.L.; Beckman, J.; Bersin, R.M.; Carr, V.F.; Casey, D.; Eagle, K.A.; Hermann, L.K.; Isselbacher, E.M.; Kazerooni, E.A.; et al. 2010 ACCF/AHA/AATS/ACR/ASA/SCA/SCAI/SIR/STS/SVM Guidelines for the Diagnosis and Management of Patients with Thoracic Aortic Disease. *J. Am. Coll. Cardiol.* **2010**, *55*, e27–e129. [CrossRef] [PubMed]
3. Members, T.F.; Erbel, R.; Aboyans, V.; Boileau, C.; Bossone, E.; di Bartolomeo, R.; Eggebrecht, H.; Evangelista, A.; Falk, V.; Frank, H.; et al. 2014 ESC Guidelines on the Diagnosis and Treatment of Aortic Diseases. *Eur. Hear. J.* **2014**, *35*, 2873–2926. [CrossRef]
4. Pape, L.A.; Tsai, T.T.; Isselbacher, E.M.; Oh, J.K.; O'Gara, P.T.; Evangelista, A.; Fattori, R.; Meinhardt, G.; Trimarchi, S.; Bossone, E.; et al. Aortic Diameter $\geq$ 5.5 Cm Is Not a Good Predictor of Type A Aortic Dissection. *Circulation* **2007**, *116*, 1120–1127. [CrossRef]
5. Heuts, S.; Adriaans, B.P.; Rylski, B.; Mihl, C.; Bekkers, S.C.A.M.; Olsthoorn, J.R.; Natour, E.; Bouman, H.; Berezowski, M.; Kosiorowska, K.; et al. Evaluating the Diagnostic Accuracy of Maximal Aortic Diameter, Length and Volume for Prediction of Aortic Dissection. *Heart* **2020**, *106*, 892–897. [CrossRef]
6. Glower, D.D. Indications for Ascending Aortic Replacement: Size Alone Is Not Enough. *J. Am. Coll. Cardiol.* **2011**, *58*, 585–586. [CrossRef]
7. Van Hout, M.; Juffermans, J.; Lamb, H.; Kröner, E.; Boogaard, P.V.D.; Schalij, M.; Dekkers, I.; Scholte, A.; Westenberg, J. Ascending Aorta Curvature and Flow Displacement Are Associated with Accelerated Aortic Growth at Long-Term Follow-Up: A MRI Study in Marfan and Thoracic Aortic Aneurysm Patients. *IJC Hear. Vasc.* **2021**, *38*, 100926. [CrossRef]
8. Korpela, T.; Kauhanen, S.P.; Kariniemi, E.; Saari, P.; Liimatainen, T.; Jaakkola, P.; Vanninen, R.; Hedman, M. Flow Displacement and Decreased Wall Shear Stress Might Be Associated with the Growth Rate of an Ascending Aortic Dilatation. *Eur. J. Cardio-Thoracic Surg.* **2021**, *61*, 395–402. [CrossRef]

9. Ramaekers, M.J.; Adriaans, B.P.; Juffermans, J.F.; van Assen, H.C.; Bekkers, S.C.; Scholte, A.J.; Kenjeres, S.; Lamb, H.J.; Wildberger, J.E.; Westenberg, J.J.; et al. Characterization of Ascending Aortic Flow in Patients with Degenerative Aneurysms. *Investig. Radiol.* **2021**, *56*, 494–500. [CrossRef]
10. Bm, S.P.K.; Hedman, M.; Kariniemi, E.; Jaakkola, P.; Vanninen, R.; Saari, P.; Liimatainen, T. Aortic Dilatation Associates with Flow Displacement and Increased Circumferential Wall Shear Stress in Patients Without Aortic Stenosis: A Prospective Clinical Study. *J. Magn. Reson. Imaging* **2019**, *50*, 136–145. [CrossRef]
11. Dux-Santoy, L.; Guala, A.; Teixidó-Turà, G.; Ruiz-Muñoz, A.; Maldonado, G.; Villalva, N.; Galian, L.; Valente, F.; Gutiérrez, L.; González-Alujas, T.; et al. Increased Rotational Flow in the Proximal Aortic Arch Is Associated With Its Dilation in Bicuspid Aortic Valve Disease. *Eur. Heart J. Cardiovasc. Imaging* **2019**, *20*, 1407–1417. [CrossRef] [PubMed]
12. Bissell, M.M.; Hess, A.T.; Biasiolli, L.; Glaze, S.J.; Loudon, M.; Pitcher, A.; Davis, A.; Prendergast, B.; Markl, M.; Barker, A.J.; et al. Aortic Dilation in Bicuspid Aortic Valve Disease. *Circ. Cardiovasc. Imaging* **2013**, *6*, 499–507. [CrossRef] [PubMed]
13. Garcia, J.; Barker, A.J.; Collins, J.D.; Carr, J.C.; Markl, M. Volumetric Quantification of Absolute Local Normalized Helicity in Patients with Bicuspid Aortic Valve and Aortic Dilatation. *Magn. Reson. Med.* **2017**, *78*, 689–701. [CrossRef] [PubMed]
14. Lorenz, R.; Bock, J.; Barker, A.J.; von Knobelsdorff-Brenkenhoff, F.; Wallis, W.; Korvink, J.G.; Bissell, M.M.; Schulz-Menger, J.; Markl, M. 4D Flow Magnetic Resonance Imaging in Bicuspid Aortic Valve Disease Demonstrates Altered Distribution of Aortic Blood Flow Helicity. *Magn. Reson. Med.* **2013**, *71*, 1542–1553. [CrossRef]
15. Sigovan, M.; Hope, M.D.; Dyverfeldt, P.; Saloner, D. Comparison of Four-Dimensional Flow Parameters for Quantification of Flow Eccentricity in the Ascending Aorta. *J. Magn. Reson. Imaging* **2011**, *34*, 1226–1230. [CrossRef]
16. Hope, M.D.; Sigovan, M.; Wrenn, S.J.; Saloner, D.; Dyverfeldt, P. MRI Hemodynamic Markers of Progressive Bicuspid Aortic Valve-Related Aortic Disease. *J. Magn. Reson. Imaging* **2014**, *40*, 140–145. [CrossRef]
17. Reijer, P.M.D.; Sallee, D.; Van Der Velden, P.; Zaaijer, E.R.; Parks, W.J.; Ramamurthy, S.; Robbie, T.Q.; Donati, G.; Lamphier, C.; Beekman, R.P.; et al. Hemodynamic Predictors of Aortic Dilatation in Bicuspid Aortic Valve by Velocity-Encoded Cardiovascular Magnetic Resonance. *J. Cardiovasc. Magn. Reson.* **2010**, *12*, 4. [CrossRef]
18. Geiger, J.; Hirtler, D.; Gottfried, K.; Rahman, O.; Bollache, E.; Barker, A.J.; Markl, M.; Stiller, B. Longitudinal Evaluation of Aortic Hemodynamics in Marfan Syndrome: New Insights from a 4D Flow Cardiovascular Magnetic Resonance Multi-Year Follow-Up Study. *J. Cardiovasc. Magn. Reson.* **2017**, *19*, 33. [CrossRef]
19. Minderhoud, S.C.S.; Roos-Hesselink, J.W.; Chelu, R.G.; Bons, L.R.; Hoven, A.T.V.D.; Korteland, S.-A.; Bosch, A.E.V.D.; Budde, R.P.J.; Wentzel, J.J.; Hirsch, A. Wall Shear Stress Angle Is Associated with Aortic Growth in Bicuspid Aortic Valve Patients. *Eur. Heart J. Cardiovasc. Imaging* **2022**, *22*, jeab290. [CrossRef]
20. Van Der Palen, R.L.F.; Roest, A.A.W.; Boogaard, P.J.V.D.; De Roos, A.; Blom, N.A.; Westenberg, J.J.M. Scan–Rescan Reproducibility of Segmental Aortic Wall Shear Stress As Assessed by Phase-Specific Segmentation With 4D Flow MRI in Healthy Volunteers. *Magn. Reson. Mater. Physics Biol. Med.* **2018**, *31*, 653–663. [CrossRef]
21. Markl, M.; Wallis, W.; Harloff, A. Reproducibility of Flow and Wall Shear Stress Analysis Using Flow-Sensitive Four-Dimensional MRI. *J. Magn. Reson. Imaging* **2011**, *33*, 988–994. [CrossRef] [PubMed]
22. Van Ooij, P.; Powell, A.L.; Potters, W.V.; Carr, J.C.; Markl, M.; Barker, J.A. Reproducibility and interobserver variability of systolic blood flow velocity and 3D wall shear stress derived from 4D flow MRI in the healthy aorta. *J. Magn. Reson. Imaging* **2016**, *43*, 236–248. [CrossRef] [PubMed]
23. Dyverfeldt, P.; Bissell, M.; Barker, A.J.; Bolger, A.F.; Carlhäll, C.-J.; Ebbers, T.; Francios, C.J.; Frydrychowicz, A.; Geiger, J.; Giese, D.; et al. 4D Flow Cardiovascular Magnetic Resonance Consensus Statement. *J. Cardiovasc. Magn. Reson.* **2015**, *17*, 72. [CrossRef]
24. Ms, S.P.; Dyverfeldt, P.; Ebbers, T. Assessment of the Accuracy of MRI Wall Shear Stress Estimation Using Numerical Simulations. *J. Magn. Reson. Imaging* **2012**, *36*, 128–138. [CrossRef]
25. Juffermans, J.F.; Westenberg, J.J.; Boogaard, P.J.; Roest, A.A.; van Assen, H.C.; van der Palen, R.L.; Lamb, H.J. Reproducibility of Aorta Segmentation on 4D Flow MRI in Healthy Volunteers. *J. Magn. Reson. Imaging* **2021**, *53*, 1268–1279. [CrossRef] [PubMed]
26. Wentland, A.L. Editorial for "Reproducibility of Aorta Segmentation on 4D Flow MRI in Healthy Volunteers. *J. Magn. Reson. Imaging* **2021**, *53*, 1280–1281. [CrossRef]
27. Della Corte, A.; Bancone, C.; Dialetto, G.; Covino, F.E.; Manduca, S.; Montibello, M.V.; De Feo, M.; Buonocore, M.; Nappi, G. The Ascending Aorta with Bicuspid Aortic Valve: A Phenotypic Classification with Potential Prognostic Significance. *Eur. J. Cardio Thorac. Surg.* **2014**, *46*, 240–247. [CrossRef]
28. Virtanen, P.; Gommers, R.; Oliphant, T.E.; Haberland, M.; Reddy, T.; Cournapeau, D.; Burovski, E.; Peterson, P.; Weckesser, W.; Bright, J.; et al. SciPy 1.0: Fundamental Algorithms for Scientific Computing in Python. *Nat. Methods* **2020**, *17*, 261–272. [CrossRef]
29. Bland, J.M.; Altman, D.G. Statistical Methods for Assessing Agreement Between Two Methods of Clinical Measurement. *Lancet* **1986**, *1*, 307–310. [CrossRef]
30. Meierhofer, C.; Schneider, E.P.; Lyko, C.; Hutter, A.; Martinoff, S.; Markl, M.; Hager, A.; Hess, J.; Stern, H.; Fratz, S. Wall Shear Stress and Flow Patterns in the Ascending Aorta in Patients with Bicuspid Aortic Valves Differ Significantly from Tricuspid Aortic Valves: A Prospective Study. *Eur. Heart J. Cardiovasc. Imaging* **2012**, *14*, 797–804. [CrossRef]
31. Schnell, S.; Rose, M.J.; Wu, C.; García, J.; Robinson, J.D.; Markl, M.; Rigsby, C.K. Improved Assessment of Aortic Hemodynamics by K-T Accelerated Dual-Venc 4D Flow MRI in Pediatric Patients. *J. Cardiovasc. Magn. Reson.* **2016**, *18*, O96. [CrossRef]

32. Ma, L.E.; Markl, M.; Chow, K.; Vali, A.; Wu, C.; Schnell, S. Efficient triple-VENC phase-Contrast MRI for Improved Velocity Dynamic Range. *Magn. Reson. Med.* **2020**, *83*, 505–520. [CrossRef] [PubMed]
33. Bons, L.R.; Duijnhouwer, A.L.; Boccalini, S.; Hoven, A.T.V.D.; van der Vlugt, M.J.; Chelu, R.G.; McGhie, J.S.; Kardys, I.; Bosch, A.E.V.D.; Siebelink, H.-M.J.; et al. Intermodality Variation of Aortic Dimensions: How, Where and When to Measure the Ascending Aorta. *Int. J. Cardiol.* **2019**, *276*, 230–235. [CrossRef] [PubMed]
34. Bell, V.; Mitchell, W.A.; Sigurðsson, S.; Westenberg, J.J.M.; Gotal, J.D.; Torjesen, A.A.; Aspelund, T.; Launer, L.J.; de Roos, A.; Gudnason, V.; et al. Longitudinal and Circumferential Strain of the Proximal Aorta. *J. Am. Heart Assoc.* **2014**, *3*, e001536. [CrossRef] [PubMed]
35. Rengier, F.; Weber, T.F.; Henninger, V.; Böckler, D.; Schumacher, H.; Kauczor, H.-U.; von Tengg-Kobligk, H. Heartbeat-Related Distension and Displacement of the Thoracic Aorta in Healthy Volunteers. *Eur. J. Radiol.* **2012**, *81*, 158–164. [CrossRef] [PubMed]
36. Kilner, P.J.; Yang, G.Z.; Mohiaddin, R.H.; Firmin, D.N.; Longmore, D.B. Helical and Retrograde Secondary Flow Patterns in the Aortic Arch Studied by Three-Directional Magnetic Resonance Velocity Mapping. *Circulation* **1993**, *88*, 2235–2247. [CrossRef]
37. Dux-Santoy, L.; Rodríguez-Palomares, J.F.; Teixidó-Turà, G.; Ruiz-Muñoz, A.; Casas, G.; Valente, F.; Servato, M.L.; Galian-Gay, L.; Gutiérrez, L.; González-Alujas, T.; et al. Registration-Based Semi-Automatic Assessment of Aortic Diameter Growth Rate from Contrast-Enhanced Computed Tomography Outperforms Manual Quantification. *Eur. Radiol.* **2021**, *32*, 1997–2009. [CrossRef]
38. Van Hout, M.J.; Scholte, A.J.; Juffermans, J.F.; Westenberg, J.J.; Zhong, L.; Zhou, X.; Schalla, S.M.; Hope, M.D.; Bremerich, J.; Kramer, C.M.; et al. How to Measure the Aorta Using MRI: A Practical Guide. *J. Magn. Reson. Imaging* **2020**, *52*, 971–977. [CrossRef]

applied sciences

MDPI

Review

Positron Emission Tomography in Coronary Heart Disease

José de Almeida [1,2,*], Sofia Martinho [1], Lino Gonçalves [1,2] and Maria Ferreira [1,2,3,*]

[1] Department of Cardiology, Coimbra Hospital and University Centre (CHUC), 3004-561 Coimbra, Portugal; sofia14martinho@gmail.com (S.M.); lgoncalv@ci.uc.pt (L.G.)
[2] Faculty of Medicine, University of Coimbra (FMUC), 3004-561 Coimbra, Portugal
[3] Institute for Nuclear Sciences Applied to Health (ICNAS), 3004-561 Coimbra, Portugal
* Correspondence: josepauloalmeida92@gmail.com (J.d.A.); mjvidigal19@gmail.com (M.F.)

Abstract: With advances in scanner technology, postprocessing techniques, and the development of novel positron emission tomography (PET) tracers, the applications of PET for the study of coronary heart disease have been gaining momentum in the last few years. Depending on the tracer and acquisition protocol, cardiac PET can be used to evaluate the atherosclerotic lesion (plaque imaging) and to assess its potential consequences—ischemic versus nonischemic (perfusion imaging) and viable versus scarred (viability imaging) myocardium. The scope of this review is to summarize the role of PET in coronary heart disease.

Keywords: positron emission tomography; coronary heart disease; plaque imaging; perfusion imaging; viability imaging; 18F-sodium fluoride; 18F-fluorodeoxyglucose; Rubidium-82; 13N-ammonia; 15O-water

Citation: de Almeida, J.; Martinho, S.; Gonçalves, L.; Ferreira, M. Positron Emission Tomography in Coronary Heart Disease. *Appl. Sci.* **2022**, *12*, 4704. https://doi.org/10.3390/app12094704

Academic Editor: Julio Garcia Flores

Received: 20 March 2022
Accepted: 4 May 2022
Published: 7 May 2022

1. Introduction

PET has been increasingly used in the medical field since its discovery, in the sixth decade of the last century, due to both technological advances and the expansion of indications. Fifty years after its development, the number of detectors increased from 64 to 19,000 and the spatial resolution of the exam from 14 mm to 4 mm [1]. These detectors measure the total energy deposited by the two annihilation photons moving in opposite directions produced after positron emission from a radionuclide-tagged tracer molecule [2]. The tracer molecule is part of the biochemical process that we are interested in quantifying and as such, the radionuclide should ideally only interact with that molecule [3]. The raw data acquired is then corrected for attenuation and finally reconstructed to provide an estimate of the in vivo tracer distribution [2].

The evolution of PET has had a huge impact in the study of cardiovascular disease, including coronary heart disease, heart failure, cardio-inflammatory disease, valvular heart disease, and assessment of cardiac devices and cardiac tumors.

Coronary heart disease is defined as an inadequate blood flow to an area of myocardial tissue due to blockage of the blood vessels that supply it, most often by an atherosclerotic plaque [4].

PET can be used to evaluate the atherosclerotic lesion (plaque imaging) and to assess its potential consequences—ischemic versus nonischemic (perfusion imaging) and viable versus scarred (viability imaging) myocardium [5].

2. Plaque Imaging

Although coronary atheromatous plaques are often present in older patients, in most cases their development is silent and will not result in a cardiovascular event. In contrast, some atherosclerotic plaques will rupture and cause myocardial infarction. Identifying these "vulnerable" plaques before they rupture and differentiating them from their "stable" counterparts is therefore a key objective for a cardiac imaging technique [6].

PET scanners have the capacity to assess any biochemical process occurring in the atherosclerotic plaque, if the appropriate surrogate tracer molecule is used.

2.1. Imaging Inflammation with 18F-Fluorodeoxyglucose (FDG)

FDG is overall the most used radioligand in PET imaging. It is a glucose analog that enters cells via facilitated glucose transporter member (GLUT) 1 and 3 and undergoes phosphorylation to become (18)F-FDG-6-phosphate, which cannot exit the cell before radioactive decay. Therefore, signal intensity correlates with cellular glucose uptake and phosphorylation, which indicates high metabolic activity. In the case of the atherosclerotic plaque, metabolic activity appears to be related to the concentration of proinflammatory macrophages, as demonstrated in animal models [7]. In humans, serum levels of myeloperoxidase were associated with carotid plaque FDG uptake [8].

Aortic FDG uptake was independently associated with cardiovascular risk factors such as increased low-density lipoprotein and total cholesterol, and with the presence of metabolic syndrome [9–11]. On average, patients with myocardial infarction (MI) were shown to have higher aortic FDG uptake compared with stable angina patients [12]. Visceral adipose tissue uptake of FDG was associated with systemic inflammatory status, and with the presence of metabolic syndrome components [13]. Additionally, it was independently associated with the severity of CAD and with the occurrence of AMI [14].

Regarding the assessment of pharmacological treatment, statins were shown to reduce the arterial FDG signal in a dose-dependent manner and pioglitazone was shown to attenuate vascular FDG uptake [15–17]. In contrast, the novel antidyslipidemia drugs Dalcetrapib (cholesteryl ester transfer protein inhibitor) and Rilapladib (lipoprotein-associated phospholipase A2 inhibitor) did not have any effect on vascular FDG activity [18,19].

There is, however, no current evidence to support the implementation of FDG-PET for atherosclerotic risk stratification in clinical practice.

2.2. Imaging Microcalcification with 18F-Sodium Fluoride (NaF)

NaF has been used to detect bone metastases and it is known to replace the hydroxyl group of hydroxylapatite in areas of calcification [20,21]. Regarding vascular calcification, two processes occur in a continuum in the atherosclerotic plaque as a healing response to the necrotic core inflammatory mediators: microcalcification, followed by macrocalcification. The stage of microcalcification renders the plaque unstable and more predisposed to rupture, making its identification an interesting surrogate marker for detecting "vulnerable" plaques. NaF-PET was shown to be able to identify microcalcification in atherosclerotic plaques in a consistently different pattern of uptake to the macroscopic calcium observed on computer tomography (CT) [22] (Figure 1).

Different groups have proposed that NaF uptake could be used as a maker of cardiovascular risk by demonstrating its correlation with various validated clinical scores for cardiovascular disease burden [23–26]. Our group showed that in a high cardiovascular risk population, NaF atherosclerotic plaque uptake was related to the burden of cardiovascular risk factors and thoracic fat volume, but there was no association between coronary uptake and calcium score [27].

In the context of acute coronary syndrome patients, the culprit plaques associated with infracted myocardium were shown to display greater NaF uptake than "non-culprit" plaques [28].

In perhaps the most significant clinical study with NaF-PET, patients with known coronary artery disease underwent NaF-PET/CT and were followed up for fatal or nonfatal myocardial infarction over 42 months. Total coronary NaF uptake predicted MI independently of age, sex, risk factors, segment involvement and coronary calcium scores, presence of coronary stents, coronary stenosis, REACH and SMART scores, the Duke coronary artery disease index, and recent myocardial infarction [29].

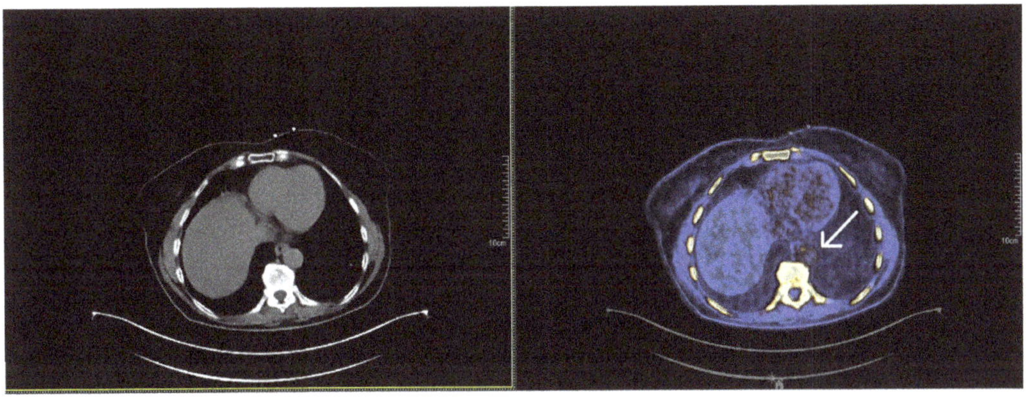

Figure 1. Fusion 18F-NaF-PET-CT images, depicting NaF uptake in the descending aorta (arrow, right picture). In the left picture, corresponding to the raw CT image, one can observe that the NaF uptake matches an area without macroscopic calcification. NaF was injected 60 min before image acquisition. Image source: Institute for Nuclear Sciences Applied to Health (ICNAS).

3. Perfusion Imaging

In perfusion imaging, stress and rest myocardial perfusion image sets are compared in order to determine the presence, extent, severity, and reversibility of stress-induced perfusion defects [30].

Similarly to single-photon emission computerized tomography (SPECT), PET can provide a visually graded qualitative assessment of relative perfusion defects (Figure 2). However, PET also allows quantitative assessment of myocardial blood flow (MBF) during the different stages of the exam, which allows the calculation of myocardial flow reserve (MFR)—the ratio of MBF in a hyperemic state and at rest (Figure 3). Quantification of absolute MBF and MFR appears to add prognostic value to the qualitative assessment [31–33].

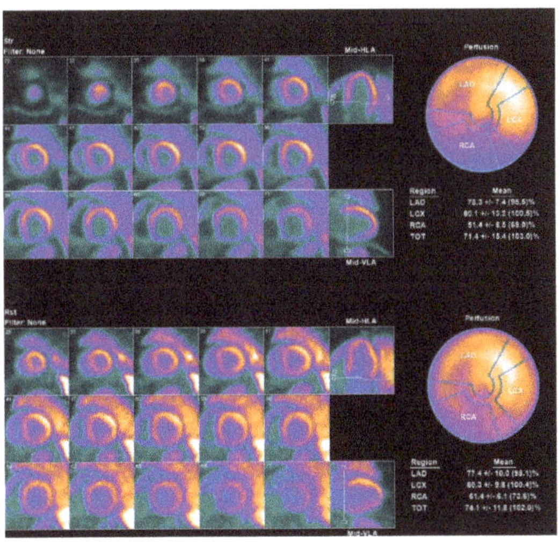

Figure 2. Assessment of myocardial perfusion with 13N-ammonia-PET. Top rows represent stress acquisition and lower rows rest acquisition. Myocardial perfusion is markedly decreased in the inferior wall during stress, compatible with ischemia in this territory. Image source: Institute for Nuclear Sciences Applied to Health (ICNAS).

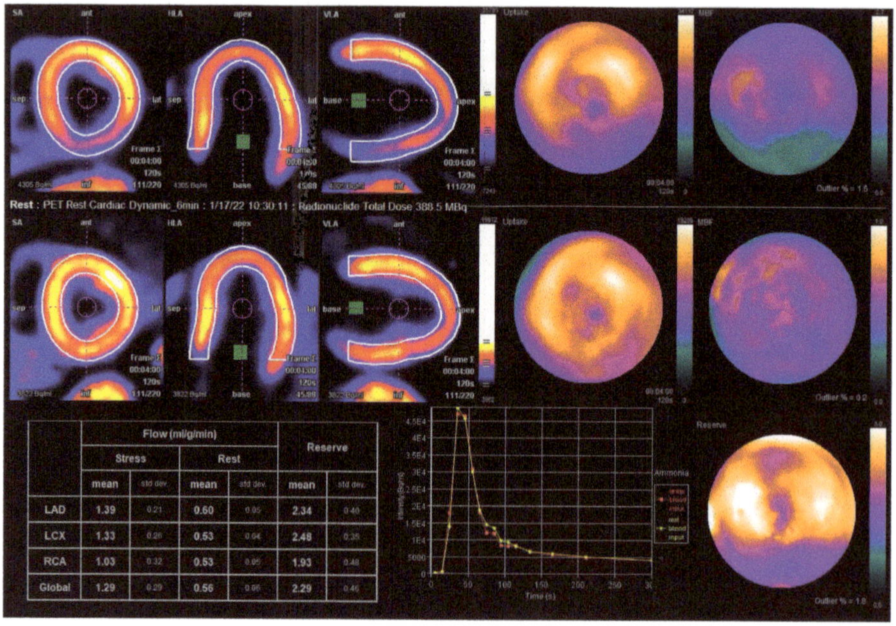

Figure 3. Quantitative assessment of myocardial perfusion with 13N-ammonia-PET of the same patient represented in Figure 2. Note that myocardial flow during stress is decreased in the RCA territory. LAD—left anterior descending artery, LCX—left circumflex artery; RCA—right coronary artery. Image source: Institute for Nuclear Sciences Applied to Health (ICNAS).

3.1. Imaging Myocardial Perfusion with Rubidium-82 (Rb)

Rb myocardial uptake is proportional to MBF, as was demonstrated for the first time more than half a century ago [34]. It is the most widespread tracer used for PET myocardial perfusion imaging (MPI), since it does not require a cyclotron on-site. It has a half-life of 78 s and an extraction fraction in comparison to MBF of around 60% [35].

Compared with conventional MPI with SPECT, Rb-PET showed improved image quality, higher diagnostic accuracy, less radiation dose to patient and staff as well as rapid examinations time [36]. It has also shown better sensitivity for the detection of multivessel disease, which in cases of balanced ischemia may present as a false negative in SPECT [35]. A recent meta-analysis compared the diagnostic performance of cardiac magnetic resonance (CMR), SPECT, and PET imaging for the identification of CAD and concluded that both CMR and PET were superior to SPECT [37]. However, a randomized study comparing the clinical effectiveness of pharmacologic SPECT and PET MPI in symptomatic CAD patients (n = 322) showed no significant differences between the two groups in subsequent rates of coronary angiography, coronary revascularization, or health status at 3-, 6-, and 12-month follow-ups [38].

In a cohort of 16,029 consecutive patients undergoing Rb rest–stress PET MPI, patients with higher degrees of ischemia had a survival benefit from early revascularization [39].

3.2. Imaging Myocardial Perfusion with 13N-Ammonia

13N-ammonia is uptaken by the cardiomyocytes, after which it is irreversibly trapped inside the cell. It has a half-life of 9.8 min and an extraction fraction in comparison to MBF of around 80% [35].

There are no studies that directly compare the diagnostic accuracy of 13N-ammonia PET with SPECT, although it may have higher sensitivity relative to Rb due to its higher myocardial extraction [35]. In unselected patients with indication for MPI, cardiac perfusion findings in 13N-ammonia PET were strong predictors of long-term outcome [40].

3.3. Imaging Myocardial Perfusion with 15O-Water

Oxygen-15-labelled water is a freely diffusible and metabolically inert tracer, and is considered the best tracer for quantitative studies [41]. It has a half-life of 2.4 min and an extraction fraction in comparison to MBF of around 95% [35].

A prospective clinical study involving 208 patients with suspected CAD who underwent CCTA, technetium 99 m/tetrofosmin–labeled SPECT, and 15O-water PET with examination of all coronary arteries by fractional flow reserve, revealed that PET exhibits the highest accuracy for diagnosis of myocardial ischemia [42]. Although not FDA approved and mainly used in research, routine clinical use of 15O-water PET with a bedside generator and infusion solution has proven to be reliable and efficient [43].

4. Viability Imaging

As previously described, one important objective of perfusion imaging is differentiating between ischemic myocardial tissue and myocardial scar. This task may be challenging with conventional perfusion imaging in the presence of myocardial hibernation. Hibernated myocardial tissue is in a state of metabolic downregulation in response to chronic or repetitive ischemia that can potentially be recovered with coronary revascularization [30].

Imaging Myocardial Viability with 18F-Fluorodeoxyglucose (FDG)

The biochemical properties of this tracer have been previously described. Demonstration of preserved glucose metabolism by FDG is a marker of myocardial viability. While reduced perfusion combined with reduced glucose metabolism suggests scarred myocardium, reduced perfusion combined with preserved or increased metabolism (mismatch) suggests hibernating myocardium.

From a theoretical point of view, management of hibernated myocardial tissue should be straightforward—revascularization. However, some studies have questioned this assumption.

In the PARR-2 (PET and recovery following revascularization) trial, patients with severe left ventricular dysfunction were randomized to revascularization decision managed by FDG-PET versus standard care. Management by FDG-PET did not result in reduction of death, MI, or recurrent hospital stay at 1 year compared with standard management [44]. However, in a post hoc analysis of a group of patients belonging to a more experienced center with ready access to FDG-PET and integration with imaging, heart failure, and revascularization teams, a significant reduction in cardiac events was observed in patients with FDG-PET-assisted management [45].

In the viability sub study of the STICH (surgical treatment for ischemic heart failure) trial, there was no significant association between myocardial viability and outcome on multivariable analysis [46].

In light of the described evidence, the most recent European Guidelines on myocardial revascularization give the use of noninvasive stress imaging for the assessment of myocardial ischemia and viability in patients with heart failure and coronary heart disease before the decision on revascularization a Class IIB recommendation [47].

5. Final Remarks

PET is a powerful tool for the diagnosis of coronary heart disease. Compared with the most commonly used nuclear exam, SPECT, it has a higher diagnostic accuracy and the addition of quantitative information yields incremental prognostic value. Cardiac PET can comprehensively assess all aspects of coronary heart disease, from coronary atherosclerotic plaque to the myocardial tissue characterization. However, how this information can be transferred to real-world practice and help to guide decision making is still a hot topic of research.

Author Contributions: Conceptualization, J.d.A.; data collection, J.d.A. and S.M.; data processing, J.d.A. and S.M.; writing—original draft preparation, J.d.A. and S.M.; writing—review and editing, L.G. and M.F. All authors have read and agreed to the published version of the manuscript.

Funding: This research received no external funding.

Institutional Review Board Statement: Not applicable.

Informed Consent Statement: Not applicable.

Data Availability Statement: Not applicable.

Conflicts of Interest: The authors declare no conflict of interest.

References

1. National Research Council (US); Institute of Medicine (US) Committee on the Mathematics and Physics of Emerging Dynamic Biomedical Imaging. *Mathematics and Physics of Emerging Biomedical Imaging*; National Academies Press (US): Washington, DC, USA, 1996.
2. Ruth, E.S.; Adam, M.A.; Paul, E.K. *The Physics of PET/CT Scanners*; Imaging Research Laboratory Department of Radiology University of Washington: Washington, DC, USA, 2013.
3. Eckelman, W.C. The application of receptor theory to receptor-binding and enzyme-binding oncologic radiopharmaceuticals. *Nucl. Med. Biol.* **1994**, *21*, 759–769. [CrossRef]
4. Criteria I of M (US) C on SSCD. *Ischemic Heart Disease*; National Academies Press (US): Washington, DC, USA, 2010.
5. Tarkin, J.M.; Ćorović, A.; Wall, C.; Gopalan, D.; Rudd, J.H. Positron emission tomography imaging in cardiovascular disease. *Heart* **2020**, *106*, 1712–1718. [CrossRef] [PubMed]
6. Andrews, J.P.M.; Fayad, Z.A.; Dweck, M.R. New methods to image unstable atherosclerotic plaques. *Atherosclerosis* **2018**, *272*, 118–128. [CrossRef] [PubMed]
7. Hyafil, F.; Cornily, J.C.; Rudd, J.H.F.; Machac, J.; Feldman, L.J.; Fayad, Z.A. Quantification of inflammation within rabbit atherosclerotic plaques using the macrophage-specific CT contrast agent N1177: A comparison with 18F-FDG PET/CT and histology. *J. Nucl. Med.* **2009**, *50*, 959–965. [CrossRef]
8. Duivenvoorden, R.; Mani, V.; Woodward, M.; Kallend, D.; Suchankova, G.; Fuster, V.; Rudd, J.H.; Tawakol, A.; Farkouh, M.E.; Fayad, Z.A. Relationship of serum inflammatory biomarkers with plaque inflammation assessed by FDG PET/CT: The dal-PLAQUE study. *JACC Cardiovasc. Imaging* **2013**, *6*, 1087–1094. [CrossRef]
9. Yoo, H.J.; Kim, S.; Hwang, S.Y.; Hong, H.C.; Choi, H.Y.; Seo, J.A.; Kim, S.G.; Kim, N.H.; Choi, D.S.; Baik, S.H.; et al. Vascular inflammation in metabolically abnormal but normal-weight and metabolically healthy obese individuals analyzed with 18F-fluorodeoxyglucose positron emission tomography. *Am. J. Cardiol.* **2015**, *115*, 523–528. [CrossRef]
10. Chróinín, D.N.; Marnane, M.; Akijian, L.; Merwick, Á.; Fallon, E.; Horgan, G.; Dolan, E.; Murphy, S.; O'Rourke, K.; O'Malley, K.; et al. Serum lipids associated with inflammation-related PET-FDG uptake in symptomatic carotid plaque. *Neurology* **2014**, *82*, 1693–1699. [CrossRef]
11. Kaida, H.; Tahara, N.; Tahara, A.; Honda, A.; Nitta, Y.; Igata, S.; Ishibashi, M.; Yamagishi, S.I.; Fukumoto, Y. Positive correlation between malondialdehyde-modified low-density lipoprotein cholesterol and vascular inflammation evaluated by 18F-FDG PET/CT. *Atherosclerosis* **2014**, *237*, 404–409. [CrossRef]
12. Joshi, N.V.; Toor, I.; Shah, A.S.V.; Carruthers, K.; Vesey, A.T.; Alam, S.R.; Sills, A.; Hoo, T.Y.; Melville, A.J.; Langlands, S.P.; et al. Systemic atherosclerotic inflammation following acute myocardial infarction: Myocardial infarction begets myocardial infarction. *J. Am. Heart Assoc.* **2015**, *4*, e001956. [CrossRef]
13. Pahk, K.; Kim, E.J.; Lee, Y.J.; Kim, S.; Seo, H.S. Characterization of glucose uptake metabolism in visceral fat by 18 F-FDG PET/CT reflects inflammatory status in metabolic syndrome. *PLoS ONE* **2020**, *15*, e0228602. [CrossRef]
14. Pahk, K.; Kim, E.J.; Joung, C.; Seo, H.S.; Kim, S. Association of glucose uptake of visceral fat and acute myocardial infarction: A pilot 18F-FDG PET/CT study. *Cardiovasc. Diabetol.* **2020**, *19*, 145. [CrossRef] [PubMed]
15. Tahara, N.; Kai, H.; Ishibashi, M.; Nakaura, H.; Kaida, H.; Baba, K.; Hayabuchi, N.; Imaizumi, T. Simvastatin attenuates plaque inflammation: Evaluation by fluorodeoxyglucose positron emission tomography. *J. Am. Coll. Cardiol.* **2006**, *48*, 1825–1831. [CrossRef] [PubMed]
16. Singh, P.; Emami, H.; Subramanian, S.; Maurovich-Horvat, P.; Marincheva-Savcheva, G.; Medina, H.M.; Abdelbaky, A.; Alon, A.; Shankar, S.S.; Rudd, J.H.; et al. Coronary plaque morphology and the anti-inflammatory impact of atorvastatin: A multicenter 18f-fluorodeoxyglucose positron emission tomographic/computed tomographic study. *Circ. Cardiovasc. Imaging* **2016**, *9*, e004195. [CrossRef]
17. Yanai, H.; Adachi, H. The low-dose (7.5 mg/day) pioglitazone therapy. *J. Clin. Med. Res.* **2017**, *9*, 821–825. [CrossRef]
18. Fayad, Z.A.; Mani, V.; Woodward, M.; Kallend, D.; Abt, M.; Burgess, T.; Fuster, V.; Ballantyne, C.M.; Stein, E.A.; Tardif, J.-C.; et al. Safety and efficacy of dalcetrapib on atherosclerotic disease using novel non-invasive multimodality imaging (Dal-plaque): A randomised clinical trial. *Lancet* **2011**, *378*, 1547–1559. [CrossRef]

19. Tawakol, A.; Singh, P.; Rudd, J.; Soffer, J.; Cai, G.; Vucic, E.; Brannan, S.P.; Tarka, E.A.; Shaddinger, B.C.; Sarov-Blat, L.; et al. Effect of treatment for 12 weeks with rilapladib, a lipoprotein-associated phospholipase A2 inhibitor, on arterial inflammation as assessed with 18F-fluorodeoxyglucose-positron emission tomography imaging. *J. Am. Coll. Cardiol.* **2014**, *63*, 86–88. [CrossRef]
20. Dweck, M.R.; Chow, M.W.; Joshi, N.V.; Williams, M.C.; Jones, C.; Fletcher, A.M.; Richardson, H.; White, A.; McKillop, G.; van Beek, E.J.; et al. Coronary arterial 18F-sodium fluoride uptake: A novel marker of plaque biology. *J. Am. Coll. Cardiol.* **2012**, *59*, 1539–1548. [CrossRef] [PubMed]
21. Creager, M.D.; Hohl, T.; Hutcheson, J.D.; Moss, A.J.; Schlotter, F.; Blaser, M.C.; Park, M.-A.; Lee, L.H.; Singh, S.A.; Alcaide-Corral, C.J.; et al. 18f-fluoride signal amplification identifies microcalcifications associated with atherosclerotic plaque instability in positron emission tomography/computed tomography images. *Circ. Cardiovasc. Imaging* **2019**, *12*, e007835. [CrossRef]
22. Derlin, T.; Richter, U.; Bannas, P.; Begemann, P.; Buchert, R.; Mester, J.; Klutmann, S. Feasibility of 18F-sodium fluoride PET/CT for imaging of atherosclerotic plaque. *J. Nucl. Med.* **2010**, *51*, 862–865. [CrossRef]
23. Blomberg, B.A.; De Jong, P.A.; Thomassen, A.; Lam, M.G.E.; Vach, W.; Olsen, M.H.; Mali, W.P.T.M.; Narula, J.; Alavi, A.; Høilund-Carlsen, P.F. Thoracic aorta calcification but not inflammation is associated with increased cardiovascular disease risk: Results of the CAMONA study. *Eur. J. Nucl. Med. Mol. Imaging* **2017**, *44*, 249–258. [CrossRef]
24. Borja, A.J.; Bhattaru, A.; Rojulpote, C.; Hancin, E.C.; Detchou, D.K.; Patil, S.; Gonuguntla, K.; Karambelkar, P.; Chinta, S.; Vuthaluru, K.; et al. Association between atherosclerotic cardiovascular disease risk score estimated by pooled cohort equation and coronary plaque burden as assessed by NaF-PET/CT. *Am. J. Nucl. Med. Mol. Imaging* **2020**, *10*, 312–318.
25. Gonuguntla, K.; Rojulpote, C.; Patil, S.; Bhattaru, A.; Karambelkar, P.; Vuthaluru, K.; Raynor, W.Y.; Borja, A.J.; Zhang, V.; Werner, T.J.; et al. Utilization of NaF-PET/CT in assessing global cardiovascular calcification using CHADS2 and CHADS2-VASc scoring systems in high risk individuals for cardiovascular disease. *Am. J. Nucl. Med. Mol. Imaging* **2020**, *10*, 293–300. [PubMed]
26. Morbelli, S.; Fiz, F.; Piccardo, A.; Picori, L.; Massollo, M.; Pestarino, E.; Marini, C.; Cabria, M.; Democrito, A.; Cittadini, G.; et al. Divergent determinants of 18F-NaF uptake and visible calcium deposition in large arteries: Relationship with Framingham risk score. *Int. J. Cardiovasc. Imaging* **2014**, *30*, 439–447. [CrossRef] [PubMed]
27. de Oliveira-Santos, M.; Castelo-Branco, M.; Silva, R.; Gomes, A.; Chichorro, N.; Abrunhosa, A.; Donato, P.; de Lima, J.P.; Pego, M.; Gonçalves, L.; et al. Atherosclerotic plaque metabolism in high cardiovascular risk subjects—A subclinical atherosclerosis imaging study with 18F-NaF PET-CT. *Atherosclerosis* **2017**, *260*, 41–46. [CrossRef]
28. Joshi, N.V.; Vesey, A.T.; Williams, M.C.; Shah, A.S.V.; Calvert, P.A.; Craighead, F.H.M.; Yeoh, S.E.; Wallace, W.; Salter, D.; Fletcher, A.M.; et al. 18F-fluoride positron emission tomography for identification of ruptured and high-risk coronary atherosclerotic plaques: A prospective clinical trial. *Lancet* **2014**, *383*, 705–713. [CrossRef]
29. Kwiecinski, J.; Tzolos, E.; Adamson, P.D.; Cadet, S.; Moss, A.J.; Joshi, N.; Williams, M.C.; van Beek, E.J.; Dey, D.; Berman, D.S.; et al. Coronary 18F-Sodium Fluoride Uptake Predicts Outcomes in Patients With Coronary Artery Disease. *J. Am. Coll. Cardiol.* **2020**, *75*, 3061–3074. [CrossRef] [PubMed]
30. Santos, B.S.; Ferreira, M.J. Positron emission tomography in ischemic heart disease. *Rev. Port. Cardiol.* **2019**, *38*, 599–608. [CrossRef]
31. Murthy, V.; Naya, M.; Foster, C.R.; Hainer, J.; Gaber, M.; Di Carli, G.; Blankstein, R.; Dorbala, S.; Sitek, A.; Pencina, M.J.; et al. Improved Cardiac Risk Assessment With Noninvasive Measures of Coronary Flow Reserve. *Circulation* **2011**, *124*, 2215–2224. [CrossRef]
32. Ziadi, M.C.; Dekemp, R.A.; Williams, K.A.; Guo, A.; Chow, B.J.; Renaud, J.M.; Ruddy, T.D.; Sarveswaran, N.; Tee, R.E.; Beanlands, R.S. Impaired Myocardial Flow Reserve on Rubidium-82 Positron Emission Tomography Imaging Predicts Adverse Outcomes in Patients Assessed for Myocardial Ischemia. *J. Am. Coll. Cardiol.* **2011**, *58*, 740–748. [CrossRef]
33. Herzog, B.A.; Husmann, L.; Valenta, I.; Gaemperli, O.; Siegrist, P.T.; Tay, F.M.; Burkhard, N.; Wyss, C.A.; Kaufmann, P.A. Long-Term Prognostic Value of 13N-Ammonia Myocardial Perfusion Positron Emission Tomography: Added Value of Coronary Flow Reserve. *J. Am. Coll. Cardiol.* **2009**, *54*, 150–156. [CrossRef]
34. Love, W.D.; Burch, G.E. Influence of the Rate of Coronary Plasma Flow on the Extraction of Rb^{86} from Coronary Blood. *Circ. Res.* **1959**, *7*, 24–30. [CrossRef] [PubMed]
35. Schindler, T.H.; Schelbert, H.R.; Quercioli, A.; Dilsizian, V. Cardiac PET Imaging for the Detection and Monitoring of Coronary Artery Disease and Microvascular Health. *JACC Cardiovasc. Imaging* **2010**, *3*, 623–640. [CrossRef] [PubMed]
36. Ghotbi, A.A.; Kjær, A.; Hasbak, P. Review: Comparison of PET rubidium-82 with conventional SPECT myocardial perfusion imaging. *Clin. Physiol. Funct. Imaging* **2013**, *34*, 163–170. [CrossRef]
37. Xu, J.; Cai, F.; Geng, C.; Wang, Z.; Tang, X. Diagnostic performance of cmr, spect, and pet imaging for the identification of coronary artery disease: A meta-analysis. *Front. Cardiovasc. Med.* **2021**, *8*, 621389. [CrossRef]
38. Patel, K.K.; Al Badarin, F.; Chan, P.S.; Spertus, J.A.; Courter, S.; Kennedy, K.F.; Case, J.A.; McGhie, A.I.; Heller, G.V.; Bateman, T.M. Randomized Comparison of Clinical Effectiveness of Pharmacologic SPECT and PET MPI in Symptomatic CAD Patients. *JACC Cardiovasc. Imaging* **2019**, *12*, 1821–1831. [CrossRef]
39. Patel, K.K.; Spertus, J.A.; Chan, P.S.; Sperry, B.W.; Thompson, R.C.; Al Badarin, F.; Kennedy, K.F.; Case, J.A.; Courter, S.; Saeed, I.M.; et al. Extent of Myocardial Ischemia on Positron Emission Tomography and Survival Benefit With Early Revascularization. *J. Am. Coll. Cardiol.* **2019**, *74*, 1645–1654. [CrossRef] [PubMed]

40. Fiechter, M.; Gebhard, C.; Ghadri, J.R.; Fuchs, T.A.; Pazhenkottil, A.P.; Nkoulou, R.; Herzog, B.A.; Altorfer, U.; Gaemperli, O.; Kaufmann, P.A. Myocardial perfusion imaging with 13N-Ammonia PET is a strong predictor for outcome. *Int. J. Cardiol.* **2012**, *167*, 1023–1026. [CrossRef]
41. Maaniitty, T.; Knuuti, J.; Saraste, A. 15O-Water PET MPI: Current Status and Future Perspectives. *Semin. Nucl. Med.* **2020**, *50*, 238–247. [CrossRef]
42. Danad, I.; Raijmakers, P.G.; Driessen, R.S.; Leipsic, J.; Raju, R.; Naoum, C.; Knuuti, J.; Mäki, M.; Underwood, R.S.; Min, J.K.; et al. Comparison of Coronary CT Angiography, SPECT, PET, and Hybrid Imaging for Diagnosis of Ischemic Heart Disease Determined by Fractional Flow Reserve. *JAMA Cardiol.* **2017**, *2*, 1100–1107. [CrossRef]
43. Tolbod, L.; Gormsen, L. Routine clinical use of 15O-water cardiac PET: Initial experience with a novel automated solution. *Eur. Heart J. Cardiovasc. Imaging* **2021**, *22* (Suppl. S3), jeab111.029. [CrossRef]
44. Beanlands, R.S.; Nichol, G.; Huszti, E.; Humen, D.; Racine, N.; Freeman, M.; Gulenchyn, K.Y.; Garrard, L.; Dekemp, R.; Guo, A.; et al. F-18-Fluorodeoxyglucose Positron Emission Tomography Imaging-Assisted Management of Patients With Severe Left Ventricular Dysfunction and Suspected Coronary Disease: A Randomized, Controlled Trial (PARR-2). *J. Am. Coll. Cardiol.* **2007**, *50*, 2002–2012. [CrossRef] [PubMed]
45. Abraham, A.; Nichol, G.; Williams, K.A.; Guo, A.; deKemp, R.A.; Garrard, L.; Davies, R.A.; Duchesne, L.; Haddad, H.; Chow, B.; et al. 18F-FDG PET imaging of myocardial viability in an experienced center with access to 18F-FDG and integration with clinical management teams: The Ottawa-FIVE sub study of the PARR 2 trial. *J. Nucl. Med.* **2010**, *51*, 567–574. [CrossRef] [PubMed]
46. Bonow, R.O.; Maurer, G.; Lee, K.L.; Holly, T.A.; Binkley, P.F.; Desvigne-Nickens, P.; Drozdz, J.; Farsky, P.S.; Feldman, A.M.; Doenst, T.; et al. Myocardial Viability and Survival in Ischemic Left Ventricular Dysfunction. *N. Engl. J. Med.* **2011**, *364*, 1617–1625. [CrossRef] [PubMed]
47. Neumann, F.J.; Sousa-Uva, M.; Ahlsson, A.; Alfonso, F.; Banning, A.P.; Benedetto, U.; Byrne, R.A.; Collet, J.P.; Falk, V.; Head, S.J.; et al. 2018 ESC/EACTS guidelines on myocardial revascularization. *Eur. Heart J.* **2019**, *40*, 87–165. [CrossRef]

applied sciences

MDPI

Review

Ultrasmall Superparamagnetic Particles of Iron Oxide and Cardiac Magnetic Resonance: Novel Imaging in Everyday Conditions

Vasiliki Tsampasian [1,2], Ioannis Merinopoulos [1,2], Donnie Cameron [1,3], Pankaj Garg [1,2] and Vassilios S. Vassiliou [1,2,*]

1 Norwich Medical School, University of East Anglia, Norwich NR4 7TJ, UK; vasiliki.tsampasian@nnuh.nhs.uk (V.T.); ioannis.merinopoulos@nnuh.nhs.uk (I.M.); donnie.cameron@uea.ac.uk (D.C.); p.garg@uea.ac.uk (P.G.)
2 Department of Cardiology, Norfolk and Norwich University Hospital, Norwich NR4 7UY, UK
3 C. J. Gorter MRI Center, Department of Radiology, Leiden University Medical Centre, 2333 ZA Leiden, The Netherlands
* Correspondence: v.vassiliou@uea.ac.uk

Abstract: Myocardial inflammation has been hypothesised to be the common underlying mechanism through which several cardiovascular diseases develop and progress. Cardiac magnetic resonance (CMR) has become a powerful non-invasive tool that enables the direct visualisation of the myocardium. The emerging use of ultrasmall superparamagnetic particles of iron oxide (USPIO) and their magnetic properties is gaining a lot of research interest. USPIO-enhanced CMR can provide valuable information, as it allows for the identification of active inflammation in the myocardium, a process that has been hypothesised to be the substrate for adverse remodelling and, eventually, heart failure. In this review, we summarise the properties of USPIO and their role in cardiac magnetic resonance imaging as well as their clinical applications.

Keywords: ultrasmall superparamagnetic particles of iron oxide (USPIO); cardiac magnetic resonance (CMR); iron oxide nanoparticles

Citation: Tsampasian, V.; Merinopoulos, I.; Cameron, D.; Garg, P.; Vassiliou, V.S. Ultrasmall Superparamagnetic Particles of Iron Oxide and Cardiac Magnetic Resonance: Novel Imaging in Everyday Conditions. *Appl. Sci.* **2022**, *12*, 6913. https://doi.org/10.3390/app12146913

Academic Editor: Julio Garcia Flores

Received: 19 May 2022
Accepted: 6 July 2022
Published: 8 July 2022

1. Introduction

A deeper understanding of cardiovascular disease has shifted the research interest towards the investigation of pathophysiological processes that encompass myocardial energetics and metabolism. Naturally, cardiovascular imaging has advanced incrementally to allow the non-invasive investigation of such processes and important features beyond anatomy and morphology that determine the biological foundation behind disease pathogenesis and progression.

Myocardial inflammation has a central role in the pathogenesis of various cardiovascular diseases. Although an important mechanism of tissue healing and recovery, the persistence of inflammation may lead to pathological and adverse cardiac remodelling [1]. Therefore, it may drive pathological processes that lead to the pathogenesis of a plethora of diseases such as dilated cardiomyopathy, ischaemic cardiomyopathy, and heart failure with a reduced or preserved ejection fraction [2,3]. Macrophages have a vital role both in the initiation and maintenance of inflammation and in the tissue repair mechanisms following an acute insult [4,5]. Therefore, the early identification of macrophage activity will enable the early detection and monitoring of a variety of diseases and may allow the steering of therapeutic management strategies towards appropriate pathways.

Cardiac magnetic resonance (CMR) imaging offers significant advantages such as a high spatial resolution, tissue characterisation, and the avoidance of harmful radiation and invasive tests [6]. With the use of advanced methods and techniques, such as late gadolinium enhancement (LGE) and T1 and T2 mapping, myocardial oedema and inflammation can be detected. However, these sequences do not allow for the identification of inflammation at a cellular level. For that reason, the use of ultrasmall superparamagnetic particles of iron oxide (USPIO) has been successfully implemented in CMR studies and has so far offered a valuable means of detecting and evaluating cellular inflammation [7–9]. USPIO-enhanced CMR is a powerful imaging technique that may reveal information vital to the diagnosis and management of cardiovascular diseases.

2. Iron Oxide Nanoparticles

Iron oxide nanoparticles were first introduced as liver-specific MRI contrast agents [10–12]. Since then, however, their application has been successfully expanded in the rapidly developing field of cardiac and vascular imaging [6,7,9,13]. Iron oxide nanoparticles are naturally found in the environment and as matter in air pollution and volcanic eruptions [14]. Superparamagnetic iron oxide nanoparticles may be generated as emissions from industry and power stations, but they can also be chemically synthesised through various methods [15]. These include classical synthesis by co-precipitation, reactions in constrained environments, the polyol method, flow injection synthesis, and sonolysis [14,15]. In addition, novel methods of nanoparticle synthesis have been described, such as the generation of magnetic α-Fe_2O_3/Fe_3O_4 heterogeneous nanoparticles via a facile solution combustion process and the synthesis of iron oxide nanoparticles by the rapid inductive heating method [16,17]. Their potential applications, safety, and bioavailability are mainly determined by two of their most essential characteristics: their coating and their size [7,18].

The core of these nanoparticles comprises iron oxide, more commonly magnetite (Fe_3O_4) or maghemite (γ-Fe_2O_3), and its radius ranges between 5 and 15 nm [19,20] (Figure 1). Iron oxide itself is hydrophobic; hence, the nanoparticles are usually modified with an external coating formed of a biocompatible hydrophilic polymer such as dextran or polyethylene glycol [7,21]. This coating is essential in order to prevent the opsonisation (i.e., binding with plasma proteins) of the nanoparticles, which would unavoidably occur in the bloodstream without it [19,21]. Additionally, through electrostatic interactions, the coating also inhibits the aggregation of nanoparticles that would otherwise take place due to their hydrophobic properties [19]. Another important consideration is an excess of iron ions may lead to the overproduction of reactive oxygen species or free radicals and subsequently cause protein, lipid, and DNA damage and, finally, cellular apoptosis [22]. Thus, the other critical 'task' of the external coating is to provide a stable cover that prevents the oxidation of the iron and its subsequent harmful effects, allowing the safe use of the nanoparticles [22]. Apart from the above, the coating that surrounds the core of the nanoparticle also acts like a 'harbour' to which ligands, drug molecules, and other molecular targets can bind [6]. Taking all of these into consideration, it is obvious why molecular imaging and targeted drug delivery are some of the areas where iron oxide nanoparticles have been successfully applied [23–25].

According to their size, iron oxide particles can be grouped into four different categories: very small superparamagnetic particles of iron oxide (VSPIO, diameter < 20 nm), ultrasmall superparamagnetic particles of iron oxide (USPIO, diameter 20–50 nm), small superparamagnetic particles of iron oxide (SPIO, diameter 20–250 nm), and micro-sized particles of iron oxide (MPIO, diameter 1–8 μm) [6,7]. However, the vast majority of the proposed applications for iron oxide nanoparticles use small and ultrasmall particles with diameters of 50 nm or less [6].

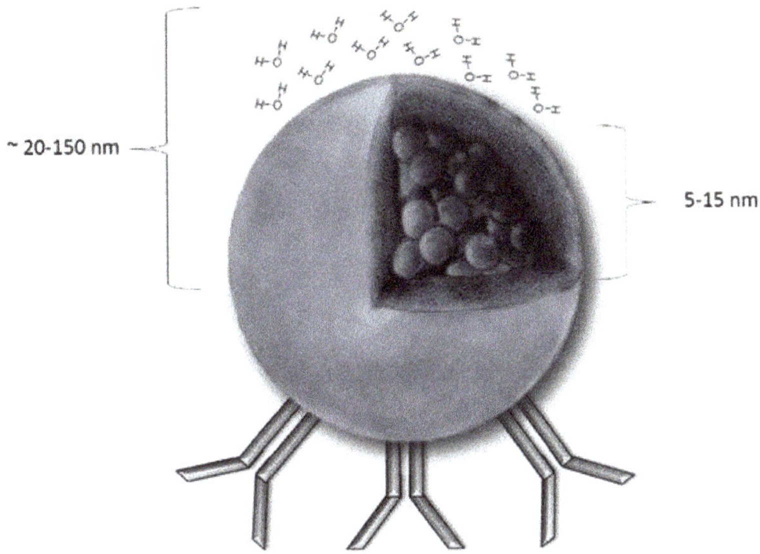

Figure 1. A schematic presentation of a SPION (superparamagnetic iron oxide nanoparticle): The core radius ranges from 5 to 15 nm, and the hydrodynamic radius (core with shell and water coat) is between 20 and 150 nm. Unless there is a magnetic field, magnetisation equals 0. As shown, SPIONs can be easily coupled with antibodies that facilitate the majority of the SPION applications discovered so far [19]. Reproduced with permission from Dulińska–Litewka et al. [19] under a Creative Commons Attribution 4.0 International License.

The concept of the superparamagnetism of nanoparticles was expressed for the first time almost a century ago [26]. The magnetic nanoparticles induce local magnetic field inhomogeneities through which water molecules will diffuse, leading to the magnetic relaxation of the water protons, from which image contrast is generated [27]. Iron oxide particles form crystalline structures that contain multiple iron ions [28]. Sufficiently large crystal-containing regions constitute a magnetic domain that comprises thermodynamically independent particles that have a net magnetic dipole that is larger than the total of the individual unpaired electrons [28]. In the absence of a magnetic field, these domains have no net magnetic field; however, in the presence of an external magnetic field, the magnetic domains are reoriented, and the net magnetic moment of the particle surpasses that of the paramagnetic ions [27,28]. This phenomenon is called superparamagnetism and appears in small ferromagnetic or ferrimagnetic nanoparticles as a result of their small size and crystalline nature, both of which are required for this phenomenon to occur [7,28,29].

3. Iron Oxide Nanoparticles (USPIO) and Inflammation

USPIO accumulate at sites of inflammation; therefore, their detection with CMR allows for the identification of active myocardial inflammation. Following intravenous administration, the iron oxide nanoparticles remain in the intravascular space and are taken up by the macrophages of the reticuloendothelial system [20]. Both the size and the coating of the nanoparticles are crucial for their bio-availability, with the smaller and more hydrophilic particles being a lot less susceptible to phagocytosis [6,23]. While larger iron oxide nanoparticles are rapidly identified and engulfed by the macrophages in the lymphoreticular organs (liver, spleen, lymph nodes), USPIO, because of their smaller size, are much less readily recognised and can subsequently avoid instant phagocytosis [6]. In this way, USPIO remain in the bloodstream for much longer, as reflected by their longer half-life (up to 36 h) compared to larger iron oxide nanoparticles (up to 2 h) [6,27]. After being taken up by macrophages, they are subsequently carried by these cells to areas of active inflammation [7]. In addition to this mechanism, it has also been suggested that iron oxide nanoparticles passively cross the endothelial barrier at sites of loss of integrity and with increased permeability [7,30]. It has been shown, however, that the main mechanism by which USPIO accumulate is predominantly via cellular infiltration within the myocardium [31].

USPIO-tagged macrophages have been identified with MRI in murine models of myocardial ischaemia and infarction [32,33]. In a proof-of-principle trial that consisted of fourteen patients with acute myocardial infarction, it was shown that USPIO have the ability to detect infiltrating macrophages, identifying areas of active inflammation [30]. In another proof-of-concept study that included sixteen patients with acute myocardial infarction, USPIO uptake was evident in areas of the infarcted and remote myocardium [34]. More recently, Lagan et al. demonstrated that USPIO can identify areas of myocardial inflammation in a murine model, with USPIO having been shown to be present inside macrophages in areas of the infarcted myocardium [1]. The same study, however, also showed that USPIO can accumulate in areas of inflammation not only after being engulfed by the macrophages but can also be 'passively' present in the interstitium outside phagocytic cells [1]. This is a significant finding, as it shows that USPIO detection is not specific for macrophage activity, and therefore, the cardiac MRI techniques used should be able to differentiate USPIO enhancement due to active macrophage uptake from the passive presence of the nanoparticles in the myocardial interstitium [1].

4. The Impact of the Iron Oxide Nanoparticles on CMR Imaging

By altering the relaxation properties of the surrounding hydrogen protons and changing the apparent proton density of the surrounding tissue, iron oxide nanoparticles influence the local magnetic resonance properties. Their use is associated with shorter T2 values and hypointense regions on T2- and T2*-weighted images [23,28]. Several factors can affect the signal intensities on T2* sequences, such as the presence of oedema or haemorrhage. For this reason, the impact of USPIO accumulation can only be evaluated with the careful examination of pre- and post- contrast images of the area of interest [13]. Typically, patients are therefore subjected to two consecutive CMR scans, with the second one being 24 h after USPIO administration [35]. More recently, it was shown that multi-time-point (at 50 and 75 h post-contrast administration) multiparametric CMR helped differentiate active USPIO uptake from macrophages from passive tissue distribution [1].

Their impact of iron oxide nanoparticles and, more specifically, USPIO can be quantified by measuring the changes in the T2* and R2* values (R2* = 1/T2*) [1,7]. Uniformly dissolved USPIO alter the relaxation rates, including the R1 (longitudinal magnetic resonance relaxation rate) and R2* (transverse magnetic resonance relaxation rate) [1,36]. Nevertheless, R2* particularly is sensitive to actively phagocytised USPIO, whereas R1 is not [1]. Therefore, the R2*/R1 ratio is extremely useful for differentiating the active USPIO uptake from the passively distributed USPIO [1,7]. In a study by Lagan et al., the ratio of R2*/R1 at 75 h post-USPIO administration was demonstrated to have a 90% sensitivity and 85% specificity for detecting active USPIO uptake in infarcted and remote myocardium in acute myocardial infarction and chronic ischaemic cardiomyopathy [1].

Iron oxide nanoparticles also cause T1 shortening [37,38]. However, the T1-relaxivity of iron oxides can vary substantially depending on the strength of the magnetic field, the size of the core of the nanoparticle, and the degree of the particle aggregation [7,20,39]. More specifically, the T1 shortening caused by the nanoparticles is diminished in areas with significant particle aggregation, with the opposite phenomenon observed in areas with diffusely located nanoparticles [40]. This, however, can be avoided through the use of lower-flip-angle radiofrequency pulses and longer repetition times [7].

5. Clinical Applications in Myocardial Imaging

Multiple studies have assessed the role of USPIO-enhanced CMR in ischaemic cardiomyopathy (Figure 2). In a study that included 16 patients with acute ST-segment elevation myocardial infarction, Alam et al. demonstrated that USPIO were taken up by the infarcted myocardial tissue and, to a lesser degree, by the peri-infarct and the remote myocardium, as evidenced by increased R2* values [34]. This was in keeping with the results of an NIMINI-2 (Non-invasive Myocardial Inflammation Imaging Based on New Molecular Magnetic Resonance Imaging) study, which showed that the absolute T2* values were decreased in the infarcted myocardial zone and, again to a lesser degree, in the peri-infarct zone [30]. In the same study, ex vivo analysis revealed that USPIO uptake was detected specifically in cultured macrophages and not in the peripheral blood monocytes, supporting the theory that USPIO absorption is mainly due to their accumulation in infiltrating myocardial macrophages or their passive accumulation in areas of loss of endothelial integrity rather than their phagocytosis by peripheral macrophages that then relocate to the heart [30].

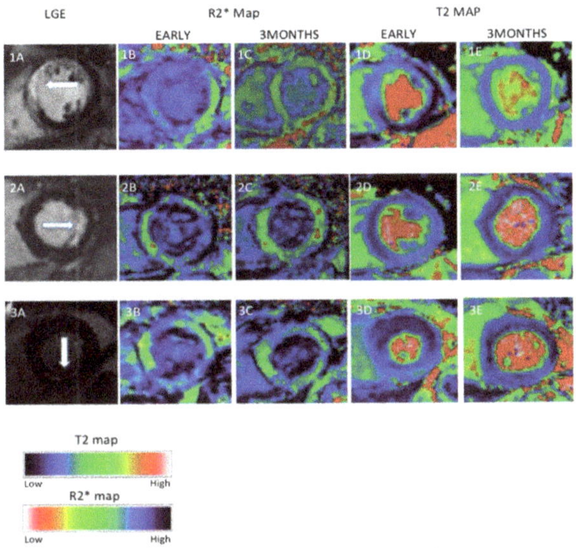

Figure 2. Examples of myocardial oedema and ferumoxytol enhancement in the infarct area after myocardial infarction (MI). Three examples of MI (1, anteroseptal, 2, lateral and 3, inferior) showing LGE on T1-weighted imaging (**1A–3A**), ferumoxytol enhancement (R2* maps) (**1B–3B,1C–3C**), and oedema (T2 maps) (**1D–3D,1E–3E**) at early (up to 10 days in (**1B–3B,1D–3D**)) and late (3 months in (**1C–3C,1E–3E**)) post-MI. Early inflammation and oedema seen on R2* maps (dark region) and T2 maps (light region), respectively, have improved or resolved by three months [31]. Reproduced with permission from Stirrat et al. [31] under a Creative Commons Attribution 4.0 International License.

These two proof-of-concept studies laid the foundation for further trials that followed and demonstrated that USPIO accumulation following acute myocardial infarction reflects active cellular inflammation in the myocardium [31,41]. As mentioned above, these studies showed that USPIO accumulated, albeit to a lesser degree, not only in the infarcted and peri-infarct but also in the remote myocardium, generating the hypothesis that inflammation and macrophage infiltration in the remote myocardium may lead to adverse myocardial remodelling. More recently, in a study that included patients with acute MI and with chronic ischaemic cardiomyopathy, Lagan et al. demonstrated that a USPIO-enhanced multi-parametric multi-time-point CMR methodology specifically differentiates active myocardial macrophage infiltration from the passive distribution of dissolved USPIO in the tissue interstitium [1]. Importantly, it was shown for the first time that there is persistent active inflammation, as evidenced by macrophage infiltration, both in the infarcted and the remote myocardium in the chronic phase following an MI [1]. Larger studies using USPIO-enhanced CMR are now needed to examine the potential role of active macrophage infiltration in the pathogenesis of adverse myocardial remodelling in ischaemic cardiomyopathy. The limited available data on USPIO uptake following coronary artery bypass graft (CABG) have demonstrated that myocardial injury following CABG may be less dependent on macrophage infiltration and inflammation, as there was no correlation between USPIO uptake and high-sensitivity cardiac troponin or the cardiopulmonary bypass time [42].

USPIO-enhanced CMR has provided essential insights into the pathophysiology of Takotsubo cardiomyopathy. While it has been known that oedema and inflammation play a central role in this disease, Scally et al. demonstrated for the first time that active macrophage infiltration drives the pathological ventricular response and cellular inflammation [43,44]. Compared to control individuals, patients with takotsubo cardiomyopathy showed significant differences in the change om T2* and native T1 values at baseline (i.e., during the acute event). However, these differences were no longer evident after five months [44]. Notably, myocardial energetics as assessed by 31P-CMR spectroscopy demonstrated a markedly reduced energetic state, which persisted after five months [44].

In contrast with takotsubo cardiomyopathy, USPIO-enhanced CMR results have not shown evidence of active macrophage infiltration and USPIO accumulation in myocarditis despite evidence of myocardial oedema in native T1 mapping and extracellular volume (ECV) imaging (Figure 3) [1,45]. Although the data available are limited to a total of fifteen patients from two studies, USPIO uptake has not been demonstrated to be substantially increased in the myocardium of patients with myocarditis, even in the areas with confirmed myocardial oedema, as shown in native T1 and ECV imaging [1,45]. This highlights the fact that different pathologies trigger different pathways that may not necessarily be mediated by macrophages. In the case of myocarditis, immune cell activation may be dominated by other cell types, such as lymphocytes, for example, which could explain the absence of USPIO accumulation in the myocardium [1,45]. Additionally, in the study by Lagan et al. there was no correlation shown between USPIO uptake and native T1 values, which, as the authors mention, emphasises that myocardial oedema and macrophage infiltration do not necessarily occur simultaneously and may not be dependent on each other [1]. Nevertheless, given the limited number of patients in the abovementioned studies, it is impossible to draw firm conclusions, and further research is warranted to investigate this important aspect in detail.

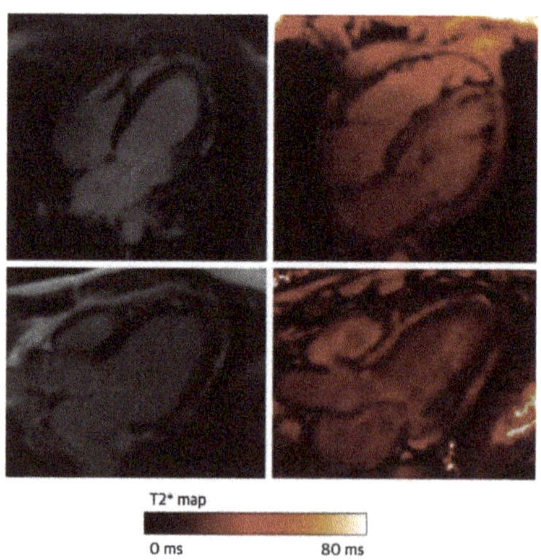

Figure 3. An example of a patient with acute myocarditis showing sub-epicardial LGE inferiorly and inferolaterally on 4-chamber and 3-chamber views (left) but no evidence of ferumoxytol uptake within the regions displaying LGE 24 h following infusion (right). CMR = cardiac magnetic resonance; LGE = late gadolinium enhancement. Reproduced with permission from Merinopoulos et al. [7].

6. Clinical Practice and Safety Profile

Ferumoxytol (Feraheme), a carboxymethyl dextran-coated USPIO, is approved for the treatment of iron deficiency anaemia in Europe and the United States. Although approved as a therapeutic agent, it is also useful as an MRI contrast agent, and as such, it is often used for research purposes under approved protocols [46].

USPIO are a safe alternative to gadolinium-based contrast agents and can even be used in patients with chronic kidney disease who are at risk of nephrogenic systemic fibrosis [47]. One of the main safety concerns is the risk of an acute hypersensitivity allergic reaction, leading the U.S. Food and Drug Administration to issue a boxed warning following 79 reported cases of anaphylaxis from an estimated total of 1.2 million injections [48]. Nevertheless, since then, large studies have successfully supported their safe use in clinical practice [46]. In a large multicentre MRI registry, in which more than 3000 patients were administered more than 4000 ferumoxytol injections, no severe or fatal adverse events occurred, while moderate adverse events were recorded in only 0.2% [47].

Over the last few years, the safety profile of USPIO has improved, and evidence from large studies further supports their application and safe use in clinical practice. Their use in CMR has successfully revealed important mechanistic pathways in which cellular inflammation and active macrophage infiltration play an important role. The research data so far reflect their successful use in the investigation of cardiac diseases and the potential that USPIO-enhanced CMR holds for use in clinical practice, to guide treatment strategies, and to be at the centre of personalised medicine.

7. Conclusions

USPIO are a safe non-invasive emerging method that can be used to evaluate myocardial inflammation by identifying active macrophage activity and infiltration. As inflammation may be an important factor driving adverse myocardial remodelling, USPIO-enhanced CMR has the potential to offer important insights into the pathophysiology of a range of cardiovascular diseases. In this way, it may become a powerful tool for risk stratification, monitoring, and targeted individualised therapeutic interventions.

Author Contributions: V.T.: original draft preparation, review and editing; I.M., review and editing; D.C., review and editing; P.G., review and editing; V.S.V., supervision, review and editing. All authors have read and agreed to the published version of the manuscript.

Funding: This research received no external funding.

Institutional Review Board Statement: Ethical review and approval were waived for this study due to it being comprehensive review of already published data.

Informed Consent Statement: Patient consent was waived as already published data were used.

Data Availability Statement: Not applicable.

Conflicts of Interest: The authors declare no conflict of interest.

References

1. Lagan, J.; Naish, J.H.; Simpson, K.; Zi, M.; Cartwright, E.J.; Foden, P.; Morris, J.; Clark, D.; Birchall, L.; Caldwell, J.; et al. Substrate for the Myocardial Inflammation–Heart Failure Hypothesis Identified Using Novel USPIO Methodology. *JACC Cardiovasc. Imaging* **2021**, *14*, 365–376. [CrossRef] [PubMed]
2. Westman, P.C.; Lipinski, M.J.; Luger, D.; Waksman, R.; Bonow, R.O.; Wu, E.; Epstein, S.E. Inflammation as a Driver of Adverse Left Ventricular Remodeling after Acute Myocardial Infarction. *J. Am. Coll. Cardiol.* **2016**, *67*, 2050–2060. [CrossRef] [PubMed]
3. Nakayama, T.; Sugano, Y.; Yokokawa, T.; Nagai, T.; Matsuyama, T.A.; Ohta-Ogo, K.; Ikeda, Y.; Ishibashi-Ueda, H.; Nakatani, T.; Ohte, N.; et al. Clinical Impact of the Presence of Macrophages in Endomyocardial Biopsies of Patients with Dilated Cardiomyopathy. *Eur. J. Heart Fail.* **2017**, *19*, 490–498. [CrossRef] [PubMed]
4. Cao, D.J. Macrophages in Cardiovascular Homeostasis and Disease. *Circulation* **2018**, *138*, 2452–2455. [CrossRef]
5. Chen, M.; Li, X.; Wang, S.; Yu, L.; Tang, J.; Zhou, S. The Role of Cardiac Macrophage and Cytokines on Ventricular Arrhythmias. *Front. Physiol.* **2020**, *11*, 1113. [CrossRef]

6. Stirrat, C.; Vesey, A.; McBride, O.; Robson, J.; Alam, S.; Wallace, W.; Semple, S.; Henriksen, P.; Newby, D. Ultra-Small Superparamagnetic Particles of Iron Oxide in Magnetic Resonance Imaging of Cardiovascular Disease. *J. Vasc. Diagn. Interv.* **2014**, *2*, 99–112. [CrossRef]
7. Merinopoulos, I.; Gunawardena, T.; Stirrat, C.; Cameron, D.; Eccleshall, S.C.; Dweck, M.R.; Newby, D.E.; Vassiliou, V.S. Diagnostic Applications of Ultrasmall Superparamagnetic Particles of Iron Oxide for Imaging Myocardial and Vascular Inflammation. *JACC Cardiovasc. Imaging* **2020**, *14*, 1249–1264. [CrossRef]
8. Tsampasian, V.; Swift, A.J.; Assadi, H.; Chowdhary, A.; Swoboda, P.; Sammut, E.; Dastidar, A.; Cabrero, J.B.; Del Val, J.R.; Nair, S.; et al. Myocardial Inflammation and Energetics by Cardiac MRI: A Review of Emerging Techniques. *BMC Med. Imaging* **2021**, *21*, 164. [CrossRef]
9. Lee, N.; Hyeon, T. Designed synthesis of uniformly sized iron oxide nanoparticles for efficient magnetic resonance imaging contrast agents. *Chem. Soc. Rev.* **2012**, *41*, 2575–2589. [CrossRef]
10. Renshaw, P.F.; Owen, C.S.; McLaughlin, A.C.; Frey, T.G.; Leigh, J.S. Ferromagnetic Contrast Agents: A New Approach. *Magn. Reson. Med.* **1986**, *3*, 217–225. [CrossRef]
11. Dias, M.H.M.; Lauterbur, P.C. Ferromagnetic Particles as Contrast Agents for Magnetic Resonance Imaging of Liver and Spleen. *Magn. Reson. Med.* **1986**, *3*, 328–330. [CrossRef]
12. Saini, S.; Stark, D.D.; Hahn, P.F.; Wittenberg, J.; Brady, T.J.; Ferrucci, J.T., Jr. Ferrite Particles: A Superparamagnetic MR Contrast Agent for the Reticuloendothelial System. *Radiology* **1987**, *162*, 211–216. [CrossRef]
13. Alam, S.R.; Stirrat, C.; Richards, J.; Mirsadraee, S.; Semple, S.I.K.; Tse, G.; Henriksen, P.; Newby, D.E. Vascular and Plaque Imaging with Ultrasmall Superparamagnetic Particles of Iron Oxide. *J. Cardiovasc. Magn. Reson.* **2015**, *17*, 83. [CrossRef]
14. Singh, N.; Jenkins, G.J.S.; Asadi, R.; Doak, S.H. Potential toxicity of superparamagnetic iron oxide nanoparticles (SPION). *Nano Rev.* **2010**, *1*, 5358. [CrossRef]
15. Wallyn, J.; Anton, N.; Vandamme, T.F. Synthesis, Principles, and Properties of Magnetite Nanoparticles for In Vivo Imaging Applications—A Review. *Pharmaceutics* **2019**, *11*, 601. [CrossRef]
16. Sharma, P.; Holliger, N.; Pfromm, P.H.; Liu, B.; Chikan, V. Size-Controlled Synthesis of Iron and Iron Oxide Nanoparticles by the Rapid Inductive Heating Method. *ACS Omega* **2020**, *5*, 19853–19860. [CrossRef]
17. Li, Y.; Wang, Z.; Liu, R. Superparamagnetic α-Fe2O3/Fe3O4 Heterogeneous Nanoparticles with Enhanced Biocompatibility. *Nanomater* **2021**, *11*, 834. [CrossRef]
18. Stirrat, C.G.; Newby, D.E.; Robson, J.M.J.; Jansen, M.A. The Use of Superparamagnetic Iron Oxide Nanoparticles to Assess Cardiac Inflammation. *Curr. Cardiovasc. Imaging Rep.* **2014**, *7*, 9263. [CrossRef]
19. Dulińska-Litewka, J.; Łazarczyk, A.; Hałubiec, P.; Szafrański, O.; Karnas, K.; Karewicz, A. Superparamagnetic Iron Oxide Nanoparticles-Current and Prospective Medical Applications. *Materials* **2019**, *12*, 617. [CrossRef]
20. Bjørnerud, A.; Johansson, L. The Utility of Superparamagnetic Contrast Agents in MRI: Theoretical Consideration and Applications in the Cardiovascular System. *NMR Biomed.* **2004**, *17*, 465–477. [CrossRef]
21. Yarjanli, Z.; Ghaedi, K.; Esmaeili, A.; Rahgozar, S.; Zarrabi, A. Iron Oxide Nanoparticles May Damage to the Neural Tissue through Iron Accumulation, Oxidative Stress, and Protein Aggregation. *BMC Neurosci.* **2017**, *18*, 51. [CrossRef] [PubMed]
22. Abakumov, M.A.; Semkina, A.S.; Skorikov, A.S.; Vishnevskiy, D.A.; Ivanova, A.V.; Mironova, E.; Davydova, G.A.; Majouga, A.G.; Chekhonin, V.P. Toxicity of Iron Oxide Nanoparticles: Size and Coating Effects. *J. Biochem. Mol. Toxicol.* **2018**, *32*, e22225. [CrossRef] [PubMed]
23. Vazquez-Prada, K.X.; Lam, J.; Kamato, D.; Xu, Z.P.; Little, P.J.; Ta, H.T. Targeted Molecular Imaging of Cardiovascular Diseases by Iron Oxide Nanoparticles. *Arterioscler. Thromb. Vasc. Biol.* **2021**, *41*, 601–613. [CrossRef] [PubMed]
24. Dikpati, A.; Madgulkar, A.R.; Kshirsagar, S.J.; Chahal, A.S. Journal of Advanced Pharmaceutical Sciences Targeted Drug Delivery to CNS Using Nanoparticles. *J. Adv. Pharm. Sci.* **2012**, *2*, 179–191.
25. Ngema, L.M.; Adeyemi, S.A.; Marimuthu, T.; Ubanako, P.; Wamwangi, D.; Choonara, Y.E. Synthesis of Novel Conjugated Linoleic Acid (CLA)-Coated Superparamagnetic Iron Oxide Nanoparticles (SPIONs) for the Delivery of Paclitaxel with Enhanced In Vitro Anti-Proliferative Activity on A549 Lung Cancer Cells. *Pharmaceutics* **2022**, *14*, 829. [CrossRef]
26. Frenkel, J.; Dorfman, J. Spontaneous and Induced Magnetisation in Ferromagnetic Bodies. *Nature* **1930**, *126*, 274–275. [CrossRef]
27. Enriquez-Navas, P.M.; Garcia-Martin, M.L. Application of Inorganic Nanoparticles for Diagnosis Based on MRI. *Front. Nanosci.* **2012**, *4*, 233–245. [CrossRef]
28. Wang, Y.X.J.; Hussain, S.M.; Krestin, G.P. Superparamagnetic Iron Oxide Contrast Agents: Physicochemical Characteristics and Applications in MR Imaging. *Eur. Radiol.* **2001**, *11*, 2319–2331. [CrossRef]
29. Sjøgren, C.E.; Briley-Sæbø, K.; Hanson, M.; Johansson, C. Magnetic characterization of iron oxides for magnetic resonance imaging. *Magn. Reson. Med.* **1994**, *31*, 268–272. [CrossRef]
30. Yilmaz, A.; Dengler, M.A.; Van Der Kuip, H.; Yildiz, H.; Rösch, S.; Klumpp, S.; Klingel, K.; Kandolf, R.; Helluy, X.; Hiller, K.H.; et al. Imaging of Myocardial Infarction Using Ultrasmall Superparamagnetic Iron Oxide Nanoparticles: A Human Study Using a Multi-Parametric Cardiovascular Magnetic Resonance Imaging Approach. *Eur. Heart J.* **2013**, *34*, 462–475. [CrossRef]
31. Stirrat, C.G.; Alam, S.R.; MacGillivray, T.J.; Gray, C.D.; Dweck, M.R.; Raftis, J.; Jenkins, W.S.A.; Wallace, W.A.; Pessotto, R.; Lim, K.H.H.; et al. Ferumoxytol-Enhanced Magnetic Resonance Imaging Assessing Inflammation after Myocardial Infarction. *Heart* **2017**, *103*, 1528–1535. [CrossRef] [PubMed]

32. Montet-Abou, K.; Daire, J.L.; Hyacinthe, J.N.; Jorge-Costa, M.; Grosdemange, K.; MacH, F.; Petri-Fink, A.; Hofmann, H.; Morel, D.R.; Vallée, J.P.; et al. In Vivo Labelling of Resting Monocytes in the Reticuloendothelial System with Fluorescent Iron Oxide Nanoparticles Prior to Injury Reveals That They Are Mobilised to Infarcted Myocardium. *Eur. Heart J.* **2010**, *31*, 1410–1420. [CrossRef] [PubMed]
33. Yang, Y.; Yang, Y.; Yanasak, N.; Schumacher, A.; Hu, T.C.C. Temporal and Noninvasive Monitoring of Inflammatory-Cell Infiltration to Myocardial Infarction Sites Using Micrometer-Sized Iron Oxide Particles. *Magn. Reson. Med.* **2010**, *63*, 33–40. [CrossRef] [PubMed]
34. Alam, S.R.; Shah, A.S.V.; Richards, J.; Lang, N.N.; Barnes, G.; Joshi, N.; MacGillivray, T.; McKillop, G.; Mirsadraee, S.; Payne, J.; et al. Ultrasmall Superparamagnetic Particles of Iron Oxide in Patients with Acute Myocardial Infarction Early Clinical Experience. *Circ. Cardiovasc. Imaging* **2012**, *5*, 559–565. [CrossRef]
35. Stirrat, C.G.; Alam, S.R.; MacGillivray, T.J.; Gray, C.D.; Forsythe, R.; Dweck, M.R.; Payne, J.R.; Prasad, S.K.; Petrie, M.C.; Gardner, R.S.; et al. Ferumoxytol-Enhanced Magnetic Resonance Imaging Methodology and Normal Values at 1.5 and 3T. *J. Cardiovasc. Magn. Reson.* **2016**, *18*, 46. [CrossRef]
36. Tanimoto, A.; Oshio, K.; Suematsu, M.; Pouliquen, D.; Stark, D.D. Relaxation Effects of Clustered Particles. *J. Magn. Reson. Imaging* **2001**, *14*, 72–77. [CrossRef]
37. Finn, J.P.; Nguyen, K.L.; Han, F.; Zhou, Z.; Salusky, I.; Ayad, I.; Hu, P. Cardiovascular MRI with Ferumoxytol. *Clin. Radiol.* **2016**, *71*, 796–806. [CrossRef]
38. Bao, Y.; Sherwood, J.A.; Sun, Z. Magnetic iron oxide nanoparticles as T1 contrast agents for magnetic resonance imaging. *J. Mater. Chem. C* **2018**, *6*, 1280–1290. [CrossRef]
39. Simon, G.H.; Bauer, J.; Saborovski, O.; Fu, Y.; Corot, C.; Wendland, M.F.; Daldrup-Link, H.E. T1 and T2 Relaxivity of Intracellular and Extracellular USPIO at 1.5T and 3T Clinical MR Scanning. *Eur. Radiol.* **2006**, *16*, 738–745. [CrossRef]
40. Taktak, S.; Sosnovik, D.; Cima, M.J.; Weissleder, R.; Josephson, L. Multiparameter Magnetic Relaxation Switch Assays. *Anal. Chem.* **2007**, *79*, 8863–8869. [CrossRef]
41. Yilmaz, A.; Rösch, S.; Klingel, K.; Kandolf, R.; Helluy, X.; Hiller, K.H.; Jakob, P.M.; Sechtem, U. Magnetic Resonance Imaging (MRI) of Inflamed Myocardium Using Iron Oxide Nanoparticles in Patients with Acute Myocardial Infarction—Preliminary Results. *Int. J. Cardiol.* **2013**, *163*, 175–182. [CrossRef]
42. Alam, S.R.; Stirrat, C.; Spath, N.; Zamvar, V.; Pessotto, R.; Dweck, M.R.; Moore, C.; Semple, S.; El-Medany, A.; Manoharan, D.; et al. Myocardial Inflammation, Injury and Infarction during on-Pump Coronary Artery Bypass Graft Surgery. *J. Cardiothorac. Surg.* **2017**, *12*, 115. [CrossRef]
43. Eitel, I.; Lücke, C.; Grothoff, M.; Sareban, M.; Schuler, G.; Thiele, H.; Gutberlet, M. Inflammation in Takotsubo Cardiomyopathy: Insights from Cardiovascular Magnetic Resonance Imaging. *Eur. Radiol.* **2010**, *20*, 422–431. [CrossRef]
44. Scally, C.; Abbas, H.; Ahearn, T.; Srinivasan, J.; Mezincescu, A.; Rudd, A.; Spath, N.; Yucel-Finn, A.; Yuecel, R.; Oldroyd, K.; et al. Myocardial and Systemic Inflammation in Acute Stress-Induced (Takotsubo) Cardiomyopathy. *Circulation* **2019**, *139*, 1581–1592. [CrossRef]
45. Stirrat, C.G.; Alam, S.R.; MacGillivray, T.J.; Gray, C.D.; Dweck, M.R.; Dibb, K.; Spath, N.; Payne, J.R.; Prasad, S.K.; Gardner, R.S.; et al. Ferumoxytol-Enhanced Magnetic Resonance Imaging in Acute Myocarditis. *Heart* **2018**, *104*, 300–305. [CrossRef]
46. Vasanawala, S.S.; Nguyen, K.L.; Hope, M.D.; Bridges, M.D.; Hope, T.A.; Reeder, S.B.; Bashir, M.R. Safety and Technique of Ferumoxytol Administration for MRI. *Magn. Reson. Med.* **2016**, *75*, 2107–2111. [CrossRef]
47. Nguyen, K.L.; Yoshida, T.; Kathuria-Prakash, N.; Zaki, I.H.; Varallyay, C.G.; Semple, S.I.; Saouaf, R.; Rigsby, C.K.; Stoumpos, S.; Whitehead, K.K.; et al. Multicenter Safety and Practice for Off-Label Diagnostic Use of Ferumoxytol in MRI. *Radiology* **2019**, *293*, 554–564. [CrossRef]
48. FDA Drug Safety Communication: FDA Strengthens Warnings and Changes Prescribing Instructions to Decrease the Risk of Serious Allergic Reactions with Anemia Drug Feraheme (Ferumoxytol) | FDA. Available online: https://www.fda.gov/drugs/drug-safety-and-availability/fda-drug-safety-communication-fda-strengthens-warnings-and-changes-prescribing-instructions-decrease (accessed on 10 April 2022).

MDPI
St. Alban-Anlage 66
4052 Basel
Switzerland
Tel. +41 61 683 77 34
Fax +41 61 302 89 18
www.mdpi.com

Applied Sciences Editorial Office
E-mail: applsci@mdpi.com
www.mdpi.com/journal/applsci